T0230654

FACILITATING
THE DEVELOPMENT
AND USE OF INTERACTIVE
LEARNING ENVIRONMENTS

Computers, Cognition, and Work
A Series Edited by
Gary M. Olson, Judith S. Olson, and Bill Curtis

Bloom/Loftin (Eds.) • Facilitating the Development and Use of Interactive Learning Environments

Bowker/Star/Turner/Gasser (Eds.) • Social Science, Technical Systems, and Cooperative Work: Beyond the Great Divide

Finn/Sellen/Wilbur (Eds.) • Video-Mediated Communication

Fox • The Human Tutorial Dialogue Project: Issues in the Design of Instructional Systems

Hoschka (Ed.) • Computers as Assistants: a New Generation of Support Systems

Koschmann (Ed.) • CSCL: Theory and Practice of an Emerging Paradigm

Moran/Carroll (Eds.) • Design Rationale: Concepts, Techniques, and Use

Opperman (Ed.) • Adaptive User Support: Ergonomic Design of Manually and Automatically Adaptable Software

Smith • Collective Intelligence in Computer-Based Collaboration

FACILITATING THE DEVELOPMENT AND USE OF INTERACTIVE LEARNING ENVIRONMENTS

Edited by

CHARLES P. BLOOM
US WEST Advanced Technologies

R. BOWEN LOFTIN
University of Houston

CRC Press
Taylor & Francis Group
Boca Raton London New York

CRC Press is an imprint of the
Taylor & Francis Group, an **informa** business

First published 1998 by Lawrence Erlbaum Associates, Inc.

Published 2020 by CRC Press
Taylor & Francis Group
6000 Broken Sound Parkway NW, Suite 300
Boca Raton, FL 33487-2742

© 1998 by Taylor & Francis Group, LLC
CRC Press is an imprint of Taylor & Francis Group, an Informa business

No claim to original U.S. Government works

ISBN 13: 978-0-8058-1850-5 (hbk)

This book contains information obtained from authentic and highly regarded sources. Reasonable efforts have been made to publish reliable data and information, but the author and publisher cannot assume responsibility for the validity of all materials or the consequences of their use. The authors and publishers have attempted to trace the copyright holders of all material reproduced in this publication and apologize to copyright holders if permission to publish in this form has not been obtained. If any copyright material has not been acknowledged please write and let us know so we may rectify in any future reprint.

Except as permitted under U.S. Copyright Law, no part of this book may be reprinted, reproduced, transmitted, or utilized in any form by any electronic, mechanical, or other means, now known or hereafter invented, including photocopying, microfilming, and recording, or in any information storage or retrieval system, without written permission from the publishers.

For permission to photocopy or use material electronically from this work, please access www.copyright.com (http://www.copyright.com/) or contact the Copyright Clearance Center, Inc. (CCC), 222 Rosewood Drive, Danvers, MA 01923, 978-750-8400. CCC is a not-for-profit organization that provides licenses and registration for a variety of users. For organizations that have been granted a photocopy license by the CCC, a separate system of payment has been arranged.

Trademark Notice: Product or corporate names may be trademarks or registered trademarks, and are used only for identification and explanation without intent to infringe.

Visit the Taylor & Francis Web site at
http://www.taylorandfrancis.com

and the CRC Press Web site at
http://www.crcpress.com

Cover design by Kathryn Houghtaling Lacey

Library of Congress Cataloging-in-Publication Data

Facilitating the development and use of interactive learning
 environments / edited by Charles P. Bloom, R. Bowen Loftin.
  p.   cm.
  Includes bibliographical references and index.
  ISBN 0-8058-1850-2 (cloth : alk. paper). — ISBN 0-8058-1851-0
 (pbk. : alk. paper).
   1. Intelligent tutoring systems. I. Bloom, Charles P.
 II. Loftin, R. Bowen.
 LB1028.73.F33 1998
  371.33′4—dc21   97-44903
                                                    CIP

CONTENTS

v

PREFACE

Early in the spring of 1993, my intelligent tutoring systems (ITSs) project team at US WEST and I were busily readying for the initial trial of our first ITS platform—LEAP—(for Learn, Explore, & Practice) with actual users in the field. To celebrate this milestone, US WEST decided to sponsor a workshop on ITSs at our research facility in Boulder, CO—a site to which we knew we had little difficulty recruiting both participants and attendees. We invited many of the leaders from the field of ITSs to participate in the workshop. In truth, our real motivation for hosting the workshop was not really to celebrate but rather to obtain some *expert feedback* about the system we had built. As I mentioned, LEAP was just about to undergo an empirical evaluation with actual end users, and we wanted to check every angle to make sure we had not forgotten something that could adversely affect the outcome, for in industry, those who fail rarely get a second chance.

In organizing the workshop, we recruited as many ITS researchers as we could who had already been through much of what we were about to experience, including representatives from academia, industry, and government laboratories. The presentations were both educational and enlightening not just as pertains to the systems that had been built but also as pertains to the many lessons that had been learned (but not widely disseminated) about the ITS development and deployment process. As the workshop progressed, the discussions evolved from talking about systems and capabilities to discussing the hazards of technology transfer. A common theme espoused by both attendees and presenters at the workshop was that ITSs were good things (surprised?). However, if we all agree that they are such good things, why are not more of them deployed and in use?

As we sat down to lunch on the second day of the workshop, we began to discuss the possibility of producing a volume to document the workshop. There was strong sentiment that the world did not need another "Here is our ITS: Isn't it wonderful" book. Rather, what we all saw as a more useful contribution was a book that recounted case histories of technology transfer, both successful and unsuccessful, as well as guides and/or methods to help ITS developers facilitate the technology transfer process. Thus, this volume was born.

As first drafts of the chapters began to come in, three separate themes emerged. The first theme centered on tools that ITS developers can employ during ITS development that should ultimately facilitate deployment. The first section of this book discusses some of these tools. Chapter 1 by Bill Clancey, titled "Developing Learning Technology in Practice," calls for a paradigmatic change in how developers view their ITSs, particularly with regard to their role in the overall scheme of education. Clancey talks about a new role for ITS researchers, that of agents of organizational change. Chapter 2 by my colleagues and myself at US WEST, most notably Scott Woff, describes the efficacy and use of quasi-experimental methods to gather design information for ITS domains where other, more empirical or knowledge-based methods are inappropriate. This chapter was motivated by a second ITS project we undertook at US WEST, that of developing an ITS for teaching object-oriented requirements modeling. What we found as we began developing this system was that the type of expertise we required to build our ITS was not readily available in our company. This led us to come up with methods for acquiring ITS knowledge that were different from the more traditional methods of knowledge acquisition. Chapter 3 by Scott Wolff of US WEST describes the work he did to get US WEST to fund development of an ITS. This chapter discusses the conduct of cost–benefits analyses to facilitate ITS funding decisions. It is interesting to note that Scott's success ultimately led to US WEST funding us to develop LEAP.

The second theme that emerged centered on case studies of both successful and unsuccessful ITS technology transfer from industry. Chapter 4 by myself, along with colleagues Scott Wolff and Brigham Bell, recounts the 3-year history of US WEST's LEAP system, including descriptions of some of the methods and strategies we used to facilitate LEAP's transfer into US WEST's operational environment. Chapter 5 by Anne McClard of US WEST describes something that seems to be sorely lacking in the ITS literature—an actual observational study of subject-matter experts as they attempt to build an ITS knowledge base from scratch. During the latter half of the second year of the LEAP project, we had actual subject-matter experts building the knowledge base with minimal developer support. McClard, an anthropologist, spent the next several months observing and interviewing LEAP's subject-matter experts and the LEAP development team to try to understand both the process of knowledge-base development and how well these two extremely different organizations of people could work together. Her chapter contains findings that should be of interest to all ITS developers. Chapter 6 by Rob Farrell and Larry Lefkowitz of Bellcore recounts the WITS project, a long-term ITS project that unfortunately met with an untimely ending, despite its many technical successes. One point the authors fail to mention in their chapter is the positive impact their work had in educating and intriguing many of the training managers from all of the telcos (including US WEST), which in turn helped us with our individual ITS efforts. Chapter 7, by Peter Bullemer and myself, from my days with Honeywell in Minneapolis, describes our attempt to establish an ITS research program by tackling the "lesser" problem of helping our corporate clients in the adoption of computer-based training technology. Although this never resulted in our receiving funding to develop an ITS the lessons we learned in both technology transfer and the management of research and development projects with multiple stakeholders from our corporation's operational environment proved invaluable in our subsequent endeavors. Chapter 8 by Bob Radlinski and Mike Atwood at NYNEX describes an ITS that was, for all intents and purposes, a major technical success but a failure in the eyes of its clients. Some of the reasons why its clients considered it a failure should prove very enlightening to future ITS developers.

The third and final theme that emerged centered on case studies of both successful and unsuccessful ITS technology transfer from government-funded R&D. Chapter 9 by my coeditor Bowen Loftin describes the long and distinguished history of ITS development at NASA. All ITS developers should take pride in the knowledge that an ITS is now a mandatory part of SpaceHab astronaut training. Chapter 10 is by the Sherlock team at the University of Pittsburgh's Learning Research and Development Center (Sandy Katz, Alan Lesgold, Ted Hughes, Dan Peters, Gary Eggan, Maria Gordin, and Linda Greenberg). This chapter describes the LRDC tutor framework that has evolved from their work on Sherlock and describes many lessons learned during the process. Chapter 11 by Jeff Norton, Julie Jones, Bill Johnson, and Brad Wiederholt of Galaxy Scientific Corporation describes their long and successful history of ITS development. Their model of ITS development is extremely pragmatic, but you cannot argue with their success.

I want to express my thanks to all of the workshop participants, including the many for whom time did not allow them to write chapters. Their contributions both during and after the workshop helped shape many of the chapters contained herein. In addition, I want to thank the many colleagues who helped in the production of this book in both word and action, particularly Hans Brunner at US WEST. Hans was my supervisor at the time we held the workshop and while we developed our ITS. It was only through his generosity that the workshop (and ultimately this book) became a reality.

Charles P. Bloom

TOOLS OF THE TRADE

1

DEVELOPING LEARNING TECHNOLOGY IN PRACTICE

William J. Clancey
Institute for Research on Learning

The US West workshop on Intelligent Tutoring Systems (ITSs)[1] examined the state of the art of instructional programs, focusing on the design and deployment of systems using Artificial Intelligence (AI) programming methods. Our objective was to appraise the progress in bringing advanced research ideas to practice and to understand the barriers and opportunities for using ITS technology in industry today.

ITSs were presented from US WEST, NYNEX, Bellcore, NASA, Galaxy Scientific, Honeywell, and the Universities of Massachusetts, Colorado, and Pittsburgh. Applications included the troubleshooting of electronic circuits, customer telecommunications operations, COBOL programming, space shuttle payload operation, introductory kinematics, cardiac arrest diagnosis and treatment, and kitchen design. These programs are distinguished from conventional computer-based training (CBT) by the use of qualitative modeling techniques to represent subject material, problem-solving procedures, interactive teaching procedures, and/or a model of the student's knowledge (Clancey, 1986, 1987, 1989; Self, 1988).

---

[1] The meeting was sponsored by US West and held in Boulder, Colorado, July 26–28, 1993. The original title of the workshop referred to ITS, so this term has been used in this chapter. Increasingly, developers have preferred to shift the focus to learning and the student, emphasize interaction, and include nontechnical tools, facilities, etc. Hence, the term "interactive learning environments" is more often used today. The focus of this paper is specifically programs developed by AI researchers, which universally were called "intelligent tutoring systems" during the decade beginning around 1977.

This chapter is based on a summary talk prepared during the meeting for presentation the afternoon of the last day to stimulate group discussion. The stance is critical and yet confident about a new beginning: As ITS research struggles in a world of more limited research funding, affordable multimedia technology makes it possible to realize the early 1970s vision of IT assistants. My primary observation is that ITS technologists have a dual objective: to develop flashy multimedia models that can be used for instruction, and to change fundamentally the practice of instructional design. This second objective—promoting organizational change—is rarely mentioned and lies outside the expertise of most academic design teams. Reflecting the maturity of the technical methods, presentations at the workshop showed an emerging interest in effective design processes for everyday business, school, and government settings. Researchers are beginning to consider how difficulties in getting people to understand and adopt ITS methods are problems of organizational change, not just technical limitations.

On the other hand, researchers understand that the greatest value of ITS technology will not be realized if AI methods are used merely to replace existing lectures or computer-based training and, consequently, evaluated solely by the standards of existing training. Indeed, many ITS designers in corporations are pressed to adopt the metrics of cost and efficiency that fit a transfer view of learning and static view of organizational knowledge. If the traditional views of learning and assessing instructional methods are applied alone, the value of ITS approaches for changing classroom instruction may not be accepted or realized.

Creatively exploiting ITS technology—to change the practice of instructional design—requires a better understanding of how models relate to human knowledge. Relating the insights of the cognitive, computational, and social sciences involves changing how scientists, corporate trainers, and managers alike think about models, work, and computer tools. Broadly speaking, models comprise simulations, subject-matter taxonomies, scientific laws, equipment operational procedures, and corporate regulations and policies. Instructional designers and developers of performance support tools must better understand how the interpretation in practice of such *descriptions* is pervaded by social concerns and values (Ehn, 1988; Floyd, 1987; Greenbaum & Kyng, 1991). Social conceptions of identity and assessment influence choices people make about what tools to use, methods for gaining information, and who should be involved in projects. Judgments about ideas reflect social allegiances, not just the technical needs of work. This broader perspective on how participation and practice relate to technology moves ITS research well beyond the original focus in the 1970s on how to create models that represent different kinds of processes in the world. If we are to inquire about what models should be created and who should create and use them, we must consider new research partnerships, new design processes, and new computational methods for facilitating rather than only automating conversations.

HOW DO WE MOVE TECHNOLOGY INTO THE MAINSTREAM?

Participants in the US West workshop experienced a striking paradox: Their instructional programs are based on methods developed over nearly 25 years in internationally known computer science and psychology research laboratories, but, effectively, no one in the multibillion-dollar industry of corporate training uses this methodology. Instead, computer-based training is barely beyond the page-turning quiz generators of the 1960s, giving all computer approaches a problematic reputation.

Technically, there is a substantial gap between academic laboratory software and most training systems used in business today. Even off-the-shelf multimedia tools are at least a decade behind ITS representation and modeling techniques. Fortunately, the movement to object-oriented or component-based commercial software provides a means for sharing tools and models, but both the technological and collaboration infrastructures are still misaligned in these two cultures: Industry is only now accepting the windows and menus interface familiar to scientists in 1980 and still views Lisp, an established tool for three decades in academia, as a foreign language. Research funding was often conceived by corporations as throwing water on someone else's garden, without establishing ways of learning new methods and perspectives (epitomized by Xerox's failure to commercialize the personal computer). Ironically, funding for AI in general contracted in the 1980s under a general complaint that the work was overhyped and not relevant to pressing problems. Such complaints bring out the real mismatch between past research and everyday business, constituting a gap in current understanding:

- Does industry understand the generality of qualitative modeling methodology to science and engineering, or is the ITS approach viewed as just a smarter page turner?
- When development costs for ITS are appraised as being too high, are the multiple uses and reusability of models considered?
- Is it surprising that ITS programs are not immediately embraced by users when participation in the projects has not included conventional instructional designers, graphic artists, workers, and managers?

There are many reasons why the ITS methods of the mid-70s are not in use today. Indeed, the reasons are overwhelming:

- The computational methods are new, a radical departure from numeric programming.
- Graphic presentations required a change in hardware and software from traditional suppliers (especially IBM).
- The use of workstations in research applications predated their availability in industry by nearly 15 years (when prices dropped by more than tenfold).
- The view of knowledge and learning in 1970s' cognitive science (and embedded in the design of ITS) is not congruent with the views of workers and managers (Nonaka, 1991).
- A "delivery" mentality for software engineering in academia and industry alike prevents a participatory relation between researchers and their sponsors.
- In the late 1980s, the workforce became more distributed, with separate business units and "integrated" (nonspecialized) employees, making centralized classroom training less appropriate.

Of all these considerations, one of the most important is the shift in how knowledge and learning are conceived. From the well-known situated-learning perspective, learning is viewed as something occurring all the time and having a tacit component (Lave & Wenger, 1991; Clancey, 1995, in press). Concepts are not merely words and definitions but ways of coordinating what we see and how we move. Understanding is spatially, perceptually, and socially embedded in activity. Activities are not merely tasks but roles, identities, and ways of choreographing interpersonal interaction. Problems are constructed by participants, not merely given. "Trouble" is defined in conversations about values, how assessments will be made, and who is participating. In these terms, documents and tools are not specified and given but open to interpretation, having new meanings and uses in changing circumstances—according to workers' experience, not merely packaged by teachers and rotely digested. Communication with coworkers is viewed as central, especially by informal relationships of friendship, developing through meetings and chance encounters (Stamps, 1997).

None of this makes the articulation of principles, rules, and policies in written text irrelevant. Rather, this situated-cognition analysis reveals how such descriptions of the world and behavior are created, in what sense they are shared, and how they are given meaning in practice. The result is that both creation of work representations, or tools, and their use must be understood in the context of work activity. Put concisely, one participant at the US West workshop said, "Classroom learning should be modeled after workplace learning, not vice versa." Crucially, we don't want to fall into an either–or mentality and impose one view on another, such as bringing CBT to the desktop or bringing on-the-job training to the classroom (indeed, this occurs!). Instead, we must ask: How do formal descriptions and training facilitate everyday recoordination and reinterpretation (Wenger, 1990)? The workplace is not just a context for learning; we are not shifting an activity from one place to another. Rather, we must reconceive what is being learned—beyond the curriculum—what problem solving is actually done on the job? Is a worker's problem to learn a rule, to interpret it, or to improvise around it? Work must be viewed systemically: How does one person's solution create a problem for someone else? Again, the shift is from formalized procedures applied in narrow, functional contexts to conversation, anticipation, broad understanding, and negotiation.

Researchers with systems in use are aware that the issues are not all technological. Broadly speaking, instructional design must include understanding and configuring interactions that occur in practice among people, systems (and tools in general), and facilities. Table 1.1 summarizes the shift from a technology-centered perspective to a view of the total system of interactions in practice.

TABLE 1.1
Shift From ITS Technology-Centered Development to Design for Everyday Use

Tool Design View	Design for Everyday Use
Technology-Centered	Practice-Centered
Teachers and students as subjects.	Users as partners in multidisciplinary design teams (participatory design).
Delivering a program in a computer box.	Total system perspective, designing the context of use: organization, facilities, and information processing.
Promoting research interests.	Providing cost-effective solutions for real problems.
Automating human roles (teacher in a box) (represent what's routine).	Facilitating conversations between people (assistance for detecting and resolving trouble).
Knowledge as formalized subject matter (models & vocabulary); learning as transfer to an individual.	Promoting everyday learning: conflict resolution and interpretation of policies (social construction of knowledge).
Transparency as an objective property of data structures or graphic designs.	Transparency and ease of use as a relation between an artifact or representation and a community of practice.

Again, we need a "both–and" mentality: The perspective on the left side is not irrelevant; it is useful and important. For example, technologists must eventually deliver something to their clients, but when the perspective on the left dominates, alternative designs are not considered, and the requirements for everyday use may be misunderstood. To a substantial degree, participatory design is spreading to the United States from Scandinavia (Bradley, 1989; Corbett, Rasmussen, & Rauner, 1991; Ehn, 1988; Floyd, 1987; Greenbaum & Kyng, 1991), but the understanding of qualitative modeling and the relation of tools to activities and knowledge is not well understood.

Interacting Activities When Using Tools

Implicit in the design perspectives summarized by Table 1.1 is a fundamental epistemological shift: not viewing models as equivalent to human knowledge. To understand this, we must better understand how models are used in practice. For example, the research of Gal (1991), in collaboration with Schön, examined how designers relate direct physical experience with materials, use of simulation models, and reinterpretation of what representations in the model mean (Fig. 1.1). For instance, in the Janus example presented by Fischer, Lemke, Mastaglio, and Morch (1991), concepts such as "work triangle" in the kitchen redesign computer assistant are related differently to other constraints and, hence, given new meaning by different clients. Rather than seeing a model as replacing encounters with physical stuff in the world or replicating human conceptual structure, we view the combination as a triad: physical being-in-world (creating information by interacting), (simulation) modeling, and conceptualizing. In particular, we must be careful to distinguish using a system such as Janus versus standing in a kitchen mock-up. Similarly, we distinguish using Woolf's cardiac arrest tutor (Woolf & Hall, 1995) and perceiving a patient in the world, moving around spatially, interacting with other people.

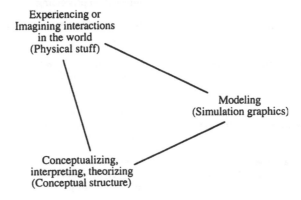

FIG. 1.1. Relating models to conceptual knowledge and activity.

The issue is not merely fidelity of isolated information—as this distinction was conceived in earlier ITS work—but the difference between interacting with a program and the choreography of being engaged in multiple activities at one time. For example, a person using a kitchen in the world is not merely "cooking" but (perhaps) "preparing a meal for a family," "catering a dinner," or "carrying on a conversation with friends." These broader aspects of human activity are nearly always excluded from functional views of places and people, which view action in terms of separated tasks defined in terms of inputs and outputs. Tasks are either serial or parallel independent; activities are conceptually coordinated simultaneously. The context of multiple activities and identities makes learning to fulfill a role more than learning scientific facts or operating procedures. In this respect, using simulation tools to replace other sources of information is far too narrow and may distort the context of practice. Effective use of simulations requires understanding how an experienced practitioner relates the constraints of actual practice (and stuff in the world) with different ways of coordinating activity (visually, in gestures, by description, rhythmically). This is an interactional view of tools—understood as being part of the relation between knowledge and activity, not a replacement or emulator for either.

Put another way, models—including, especially, ITS tools—are appropriately integrated into several interacting activities. In the domain of kitchen design this includes:

- Projecting future interactions via imagination and trials with a physical mock-up (includes role-playing and pretending).
- Modeling (abstractly representing the situation, such as creating a diagram in Janus).
- Conceptualizing (especially creating new vocabularies for describing and resolving conflicting constraints such as understanding the meaning of "work triangle").

This analysis reveals that a distinction must be drawn between learning about science and learning to design. Design inherently involves participation, values, and human activity. Designing means relating to clients, suppliers, contractors, and the community of practice of other designers, not merely learning principles for good design (Schön, 1987). Tools for design, therefore, could extend beyond a "book" view of design to supporting conversations with other people involved in the design or building process.[2] The same observations hold for troubleshooting, control, auditing, and other engineering and business activities.

---

[2]A major effort to transform classroom learning for children, the MMAP project at IRL, conveys mathematical concepts in the context of design activities (Moschkovich, 1994).

The scope of early ITS research limited its value for real-world problem solving. Knowledge was viewed as scientific, objective, and descriptive. The choice of subject material to model reveals this school-oriented, textbook bias: mathematics, physics, and electronic troubleshooting. Even most work on medical diagnosis focuses only on descriptive concepts and presumes that an *individual* (the budding "expert") is in control. Both the *epistemology* (human knowledge equals descriptive models) and the *organizational perspective* (experts manage tasks) fail to fit the reality of practice. Few ITSs orient students to other people in the setting who might be involved in framing the problem and assisting. In short, situated learning means interacting within the physical and social context of everyday life, not sitting alone in front of a screen as a disembodied, asocial learner. Again, the point of this analysis is not to rule out simulations of teams (such as an emergency room) or complex physical devices (such as a flight simulator). The point is to understand how a simulation relates to the many ways in which people coordinate their behavior and how rehearsal relates to being on the scene, working every day.

TOWARD A DYNAMIC VIEW OF REPRESENTATIONAL TOOL CREATION AND USE

To understand better how models relate to knowledge and activity, we should consider that models are not merely scientific or derived by scientific experimentation. Models of work include:

- Rules of thumb (perhaps formalized by an expert in a different culture).
- Regulations (global standards, prone to change and/or highly complex).
- Constraints articulated from interactional experience (personal values).

Understanding that professional knowledge is not merely scientific (let alone exclusively descriptive) is part of the realization that learning technology must not only move from the classroom to the workplace but must also be based on a reinvestigation of what needs to be learned. Where do design rationales come from and how are they referred to and used within a community of practice over time? A student is not only learning theories and principles but also new ways of seeing and coordinating action, new ways of talking to other people, and new interpretations of practical constraints.

In effect, instructional designers need a dynamic view of how documents and tools are modified, reinterpreted, and used to create and understand systems in the world. Usually left out of ITS evaluations are the user's experiences and reconceptualizations, which are inherently outside the formalized model of subject material. They are outside because *using* a model inherently means relating it and interpreting it in a dynamic, interactive context of activity, concern, and conflict that the model itself does not represent (Kling, 1991). Even when an ITS is placed on the job, the specification of what it must do leaves out the user's conception of identity and matrix of participation with other people over space and time.

Perhaps the best way to break out of the idea that human knowledge equals descriptive models is to contrast the curriculum orientation of pretesting and posttesting and concern with retention with views of learning based on the group's capability. How could an ITS facilitate handling unusual and difficult situations, which are both dynamic and team oriented? How could an ITS promote organizational innovation and competitiveness? Narrow, individual transfer views of training ignore how ITSs might transform organizational learning.

BRINGING TOOLS TO PRACTICE REQUIRES
ADDITIONAL RESEARCH AND NEW PARTNERS

Understanding new opportunities for learning technology—broadening our view of how ITS tools might be used in practice—we need new partners with new methods for relating designs to practice. In the past, and to some degree appropriately, ITS researchers have adopted the technical perspective of improving the capability of a computer to automate human actions. The two questions generally asked are:

1. How can we represent a design or problem situation (e.g., layout representation for a kitchen, simulation of a phone menu)?
2. How can we provide instructional feedback (e.g., critiquing and examples)?

Research has focused on methods for teaching concepts (descriptions of the world) and what to do, focusing on local constraints. The methods include qualitative modeling, natural language generation and recognition, graphics, and video.

In contrast, a practice perspective—promoting organizational change and customer orientation—asks different questions:

1. Who can benefit from such a tool? (e.g., experienced designers use the tool for *developing* theories instead of *delivering* models to novice users).
2. How can we learn about the appropriateness of the product being designed in context? (e.g., How do Janus users determine their clients'

needs? How do ITS designers determine that voice dialogues are appropriate for ordering pizzas?)

Research on design in practice has focused on methods for determining how people actually do their work and how their preferences develop through interpersonal interaction. Ironically, as with ITSs, this research is concerned about the "mental models" of users, but it considered, instead, people's views of how their jobs relate to the overall organization, their attitude about other people, and their knowledge of other people's capabilities (Levine & Moreland, 1991). Again, expert knowledge is not just about scientific theory but *about other people*. Ethnographic studies are now applying this perspective to the design of work systems, including organization, technology, and facilities (Kukla, Clemens, Morse, & Cash, 1992; Scribner & Sachs, 1991).

Helping students understand interactions in practice means relating tools to the context in which they will be used. Fact, law, and procedure views of knowledge tend to leave out this aspect of work—articulating the problem situation, creating new representations in practice. Just as ITS designers leave value judgments out of the knowledge base to be taught, they don't consider how ITS technology itself should be appropriately used. As stated by Alfred Kyle: "We know how to teach people how to build ships, but not how to figure out what ships to build" (Schön, 1987, p. 11).

Realizing this vision of designing learning technology for practice requires new research partnerships. Don Norman (1991) summarized the process neatly in the title of his position paper, "Collaborative Computing: Collaboration First, Computing Second." The point is that computer scientists need help to understand how models are created and used in practice. The design process must include social scientists who are interested in relating their descriptive study techniques to the constraints of practical design projects:

> Building applications which directly impact people is very different from building computational products of the sort taught and studied in standard courses on computer science.... (p. 88)

> Technology will only succeed if the people and the activities are very well understood.... (p. 89)

> Computer scientists cannot become social scientists overnight, nor should they.... The design of systems for cooperative work requires cooperative design teams, consisting of computer scientists, cognitive and social scientists, and representatives from the user community. (p. 90)

How shall we design the process of ITS design? Probably the simplest starting point in setting up an ITS project is to consider who the stakeholders are and how they will interact with each other during the project. People who might be involved include:

- Customers of the company's product or service.
- Workers, the company's product designers, and service providers.
- Unions.
- Managers (supervisors, teachers).
- The tool shell and application designers and researchers.
- Corporate or government "process owners."

Important questions to be negotiated as ways of working together include:

- How are conflicting goals, values, and interpretations represented and reconciled?
- Who is allowed to develop, update, or customize the tool?
- What is the division of labor in system building (teacher, knowledge engineer, programmer) and during instructional interactions?
- How are vocabularies and models shared and reused?

In considering these questions, we must remember whether something useful, durable, and scaleable will be developed. When done well, the descriptive modeling aspect of ITS technology promotes a new kind of interplay between theory and practice, requiring people to articulate and reflect on their standards, values, and experience. This interplay is missing from the "delivery" or "capture-and-disseminate" model of technology development.

As another step, ITS researchers engaged in designing for practice find it useful to report their work in a different way. Rather than only show representations and subject-matter course examples, research presentations comment on generality, the development and formalization processes, and the context of use (illustrated by the articles about Leap and Sherlock by Bloom & Lesgold in this volume). Each of these considerations is another area for research, changing traditional views about the nature of software engineering and the expertise of computer scientists. Reflecting this shift in concern, claims about instructional tools might be organized as follows:

Applying modeling methods to different domains and tasks
- Does the tool provide a general model of some domain, situation-specific models, or only a calculus for formulating models?
- What assumptions are made about systematicity or completeness of models?

Supporting development and maintenance
- How will authors, teachers, and students learn to use the program?
- What customization is possible for local needs?
- What provision is made for breakdowns in the computer tool itself?

Formalizing authoritative knowledge
- How are local constraints related to global standards?
- Who interprets regulations and policies when creating examples?
- What is the role of corporate training relative to local trainers?

Integrating technology with the context of use
- How is the tool related to existing physical, technical, organizational systems?
- How does learning already successfully occur without the tool?
- How does the tool relate to ongoing experience on the job?

Such a decomposition of concerns suggests shifting how a tool is presented at conferences and in print. Rather than describe representational techniques alone, developers working in the context of use need to discuss the system that surrounds the tool, including competing methods for learning and the practices that make a tool successful.

TRANSFORMING INSTRUCTIONAL PRACTICE: ASSESSING BOTH TECHNICAL AND SOCIAL CONCERNS

Many problems arise in developing a computer tool in practice. Until recently, ITS researchers have been preoccupied by technical concerns and have not incorporated methods and partners for handling social concerns. In turn, omitting social concerns imposes a narrow view of evaluation of ITS tools, which, among other things, inherently devalues the possible implications of this technology on organizational learning. In particular, we might revise our standards of evaluation to assess how ITS technology changes the knowledge of a group.

The questions that are usually raised when evaluating an instructional program involve measurements centered on "courseware," including authoring time, coverage of the subject matter, media, cost, student time, performance, and retention. As mentioned previously, issues of generality of the tools and customizability are often raised. A technical perspective leads us to ask, "Should we push the machine's capability by including natural language input, or should we use graphic editors?" "Should we increase the depth of the material, or should we expand the audience?" The technical focus remains on how to get something useful working and then how to make it more technologically flashy, not how to bring about organizational and technological change. Despite the obvious inadequacy of such measures of success, few people question this discourse.

In the 1990s, the widespread availability and use of workstations for clerical tasks has shifted our concern: How might technology be used to help people such as administrators, who cover what were previously multiple jobs in services, technical support, and marketing? Industry has a pressing need for the training of new employees (particularly foreign); key jobs such as telephone service for customers have a high turnover, increasing training requirements. These organizational changes provide an opportunity for reframing how ITS technology is assessed: How might it be designed to support organizational change by supporting learning in everyday contexts?

By relating ITS technology to the current business fad that "one person does it all"—variously called "the integrated-process employee," "the integrated, customer-services employee," or "the vertically-loaded, customer-oriented job"—we have the opportunity to jump past the association of computer tools for learning with CBT and classroom instruction. However, integrating learning into everyday work requires seeking new forms of assessment and inventing new kinds of supporting technologies for these restructured work functions. The conventional view that solutions to the training problem will "hop from boxes" is obviously inadequate; for the schoolroom, textbook epistemology of ITS is incomplete. Designers and managers alike need to:

- Reconsider what knowledge is required, and then relate the technology to how knowledge develops and how standards are articulated and interpreted in everyday conversations between workers.
- Recognize that part of the lack of collaboration between subject-matter experts and instructional designers derives from the false expectation that an expert can articulate what he knows; failed collaborations threaten the instructional delivery view that the most important learning occurs by training.
- Recognize that not everyone wants to learn or *will be allowed to*—issues of role, power, authority, and conflicting values must be considered.

The bottom line is that ITS developers must become engaged in a critique of the current fad of "one person does it all" and the "end of specialization," and help find a middle ground that appropriately reflects what experienced telephone representatives know and how they might productively use computer systems while still interacting with each other.

Such a discussion only begins to address the reality today in relating ITS technology to practice. The reality is a world of downsizing, decentralization, and dismantling training departments (*Training*, 1994). The reality is a world of cost containment, increasing demands on workers, and increasing reliance on computer systems to store, sort, prioritize, and monitor work. On the plus side, corporate America is becoming aware that qualitative modeling techniques exist and that computers can be more than automated page turners. The possibilities of the coaching metaphor, multiagent simulations, conceptual modeling, case-based inquiry, and conversation facilitation are still on the horizon.

An ITS researcher and designer thrown into this world is faced with a fundamental issue of not just "What ship should I build?" but "How do I change the practice of instructional design?" How do we move an organization forward? The technologist might have begun by answering, "Give the workers new tools." However, there are an amazing number of alternative and complementary approaches:

- Put people together, encouraging friendship (to promote later collaboration), mimicking, and sharpening identity through contrast.
- Dismantle the organization.
- Remove control from above; promote bottom–up restructuring.
- "Incentivize" people to do something extra (e.g., relate pay to sales).
- Experiment with different designs in different groups.
- Send people to seminars.
- Politically manipulate or resist the organization.
- Quit (join a competitor, start your own company).
- Force change by managerial fiat.

In promoting new kinds of learning technology, we must remember that such concurrent influences are part of the process by which new tools are spread, used in new ways, or rejected. Of these, ITS researchers might be advised to use the "put people together" approach: Engage early adopters of the new technology in a project that aims to develop a new understanding of learning and work, involving other participants peripherally so they can advocate similar projects in their own organizations. In the jargon of organizational change, the issue is to make the methodology "scaleable," that is, to ensure that it can be understood and usefully employed by other people. Technologists generally view this in terms of better tools, but, of course, managers will assess the methods by cost effectiveness and demonstrated results.

In summary, developing learning technology in practice—bringing ITS methods to widespread use—involves multiple concerns. The scope of evaluation and perspective on value must broaden from *representing* and *automating* to *changing practice*. The old questions must be juggled with quite different concerns that other players from the social sciences, management, and the workplace will raise and, hopefully, help resolve. Stated in terms of broadening scope, these questions include how to:

- Create instructional programs with qualitative models and multimedia.
- Augment instructional material (e.g., by modeling, graphics, explanation).
- Develop courseware (e.g., help authors, reuse components, facilitate maintenance).
- Evaluate instructional value (retention or customer satisfaction); determine how to observe changed practice and inform instructional designers.
- Decide what tools to build.
- Integrate tools into existing technological, physical, and social systems that are rapidly changing (when subject-matter experts are too busy to help or for tasks that perhaps no one knows how to do).
- Handle political, administrative, and funding conflicts among management, operations, and training departments.
- Promote organizational change, innovation, and competitiveness.

Narrow, subject-matter-oriented views of evaluation—almost universally adopted by ITS researchers in their attempt to curry the favor of instructional designers—do not show training to best advantage; they don't show learning on the job as people incorporate and expand beyond the formalized subject material of the curriculum. The ITS community, in attempting to participate in training organizations, has set its sights too low by adopting the individual transfer metaphor and managerial assessments based on what is visible—the number of student hours and the size of binders.

To shift from developing technology to developing organizations, new instructional practice, and a new view of learning, we must reframe the problem around work practice. For example, viewing a community of practice spread over a corporation's offices in several states, we might ask, "How do we support a process at a distance?" We shift from a subject-matter view of facts and theories to a view of a coordination problem—scheduling one's time, allocating resources, informing others in a timely manner—the work of choreographing contributions in a distributed system. We shift from delivering someone else's model to asking, "How do we support visualizing, relating and comparing, and organizing alternatives?" We think in terms of tools to help people represent and reflect on their own work, and on their social interactions and conceptions about other people.

To develop learning technology in practice, we must involve people in projects that straddle academic disciplines. The best hope is certainly with the next generation of students, in stark contrast with the "knowledge as static repository" view that we must capture and proceduralize the specialized viewpoints of retiring experts. The next generation will learn to describe problems and situations in multiple languages by using different perspectives and methods. This is not the same as delivering a course on one subject.

The problem for people who want to promote this new perspective on knowledge and learning is to find a way to sell what is not in a binder or flashing colorful graphics and sounds in computer windows. We must develop new forms of assessing learning that make social processes visible to reveal the informal, tacit components of knowledge as essential. The worst possible step, which the ITS community has indeed adopted so far, is to view ITS technology as supplying training departments with methods to produce more materials or better materials more quickly. Changing the practice of instructional design involves learning about and incorporating new perspectives in computer programs:

- Relating the technology of training to a worker's view of learning.
- Relating the scientific view of knowledge as written facts, rules, and theories to the pragmatics of inventing designs and creatively interpreting policies.
- Relating the individual view of work and knowledge to the social conception of activity, participation, and coordination.

Anyone who has tried to bridge such perspectives knows that the process of change is slow, politically charged, and in many respects, uncontrolled. Some researchers will prefer and will be better off working alone in research labs, exploring what computers can do rather than investigating what people need them to do. Other researchers will find that it is possible to belong to multiple communities, retaining academic ties while enjoying the intellectual breadth of a multidisciplinary team. Ultimately, beyond all the arguments about productively and competitiveness, it will be individual workers and researchers alike who assess their own performances and realize that in taking on challenges from different perspectives, they have landed on two feet in multiple worlds.

ACKNOWLEDGMENTS

This research has been supported in part by NYNEX Science and Technology, Inc. and Xerox Corporation, Inc. I want to thank my colleagues for our many stimulating conversations: Gitti Jordan, Charlotte Linde, and Ted Kahn at IRL, plus Pat Sachs, Dave Torok, and David Moore at NYNEX. Jeff Kelley and others involved with The Productivity Partnership have provided useful observations about trends in corporate training.

REFERENCES

Bradley, G. (1989). *Computers and the psychosocial work environment.* London: Taylor & Francis.

Clancey, W. J. (1986). Qualitative student models. In J. F. Traub (Ed.), *Annual Review of Computer Science, 1*, 381–450.

Clancey, W. J. (1987). *Knowledge-based tutoring: The GUIDON program.* Cambridge, MA: MIT Press.

Clancey, W. J. (1989, Summer). Viewing knowledge bases as qualitative models. *IEEE Expert,* pp. 9–23.

Clancey, W. J. (1995). A tutorial on situated learning. In J. Self (Ed.), *Proceedings of the International Conference on Computers and Education (Taiwan)* (pp. 49–70). Charlottesville, VA: AACE.

Clancey, W. J. (1997). The conceptual, nondescriptive nature of knowledge, situations, and activity. In P. Feltovich, K. Ford, & R. Hoffman (Eds.), *Expertise in context* (pp. 247–291). Avebury.

Corbett, J. M., Rasmussen, L. B., & Rauner, F. (1991). *Crossing the border: The social and engineering design of computer integrated manufacturing systems.* London: Springer-Verlag.

Ehn, P. (1988). *Work-oriented design of computer artifacts.* Stockholm: Arbeslivscentrum.

Fischer, G., Lemke, A. C., Mastaglio, T., & Morch, A. I. (1991). *The role of critiquing in cooperative problem solving.* University of Colorado Technical Report. TOIS, *9*(2), 123–151.

Floyd, C. (1987). Outline of a paradigm shift in software engineering. In Bjerknes et al. (Eds.), *Computers and democracy: A Scandinavian challenge* (p. 197). Avebury.

Gal, S. (1991). *Building bridges: Design, learning, and the role of computers.* Unpublished doctoral dissertation, APSP Research Program. MIT.

Greenbaum, J., & Kyng, M. (1991). *Design at work: Cooperative design of computer systems.* Hillsdale, NJ: Lawrence Erlbaum Associates.

Kling, R. (1991). Cooperation, coordination, and control in computer-supported work. *Communications of the ACM, 34*(12), 83–88.

Kukla, C. D., Clemens, E. A., Morse, R. S., & Cash, D. (1992). Designing effective systems: A tool approach. In P. S. Adler & T. A. Winograd (Eds.), *Usability: Turning technologies into tools* (pp. 41–65). New York: Oxford University Press.

Lave, J., & Wenger, E. (1991). *Situated learning: Legitimate peripheral participation.* Cambridge, UK: Cambridge University Press.

Levine, J. M., & Moreland, R. L. (1991). Culture and socialization in work groups. In L. B. Resnick, J. M. Levine, & S. D. Teasley (Eds.), *Perspectives on socially shared cognition* (pp. 257–279). Washington, DC: American Psychological Association.

Moschkovich, J. (1994). *Assessing students' mathematical activity in the context of design projects: Defining "authentic" assessment practices.* Paper presented at the 1994 Annual Meeting of the American Educational Research Association, New Orleans.

Nonaka, I. (1991, November–December). The knowledge-creating company. *Harvard Business Review,* pp. 96–104.

Norman, D. (1991). Collaborative computing: collaboration first, computing second. *Communications of the ACM, 34*(12), 88–90.

Schön, D. A. (1987). *Educating the reflective practitioner.* San Francisco: Jossey-Bass.

Scribner, S., & Sachs, P. (1991). *Knowledge acquisition at work.* IEEE Brief. No. 2. New York: Institute on Education and the Economy, Teachers College, Columbia University. December.

Self, J. (1988). *Artificial intelligence and human learning.* London: Chapman & Hall.

Stamps, D. (1997). Communities of practice: Learning is social. Training is irrelevant? *Training, 34*(2), 34–44.

Training, (1994, May). Re-engineering the training department. *Training,* pp. 27–34.

Wenger, E. (1990). *Toward a theory of cultural transparency: Elements of a social discourse of the visible and the invisible.* Unpublished doctoral dissertation, University of California, Irvine.

Woolf, B. P., & Hall, W. (1995, May). Multimedia pedagogues: Interactive systems for teaching and learning. *IEEE Computer (Special Issue on Multimedia),* pp. 74–80.

2

Using Quasi-Experimentation to Gather Design Information for Intelligent Tutoring Systems

A. Scott Wolff
Charles P. Bloom
Anoosh K. Shahidi
Robert E. Rehder
Applied Research
US WEST Advanced Technologies

The development of any Intelligent Tutoring System (ITS) requires the acquisition of diverse and extensive design information. This design information typically includes selecting appropriate pedagogical approaches, understanding typical errors that learners exhibit, identifying critical skills and knowledge required for performance in the domain, and so on. The manner in which this design information is acquired depends on the nature of the domain.

One premise of this chapter is that there is a continuum of "formulation" on which domains lie. At one end of this continuum are *well-formulated domains* for which there exist established procedures and expertise not only in the particular task domain but also in the training of that domain. At the other end of the continuum are *poorly formulated domains* for which there is a lack of established procedures and expertise, both for the task domain and for training in the domain.

In poorly formulated domains, in addition to ensuring that student assessment, pedagogical decision-making processes, and instructional environment of the ITS are valid, the ITS designer may be required to develop sources of expertise and knowledge of the target domain. In such cases, the designer must be concerned with ensuring that the information that the ITS teaches is valid. The best approach to accomplishing this is conducting empirical investigations.

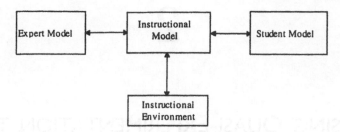

FIG. 2.1. Traditional ITS architecture.

Unfortunately, most organizations do not have the resources or appetite for randomized experimental studies that map human cognition in a domain. Additionally, ITS designers may not have the luxury to conduct such studies, given externally imposed deadlines (e.g., from clients) for system development. For many ITS designers, the key to collecting empirical data under real-world constraints is not to completely abandon experimental studies but to employ *quasi-experimental* field methods to work within organizational constraints.[1] Such methods can provide useful data to inform the design of ITSs and may even have more validity (in terms of the ITS designer's interests) than full experimental studies.

In this chapter, we first describe what specific design information is needed to build an ITS. Next, we outline several issues with regard to using empirical studies to gather design information of which the ITS designer should be aware. Following that, we present two case studies of field research conducted for different ITSs, one for teaching spreadsheet programming, the other for teaching object-oriented modeling. Finally, we present some suggestions for designing and conducting quasi-experimental field research for the purpose of designing an ITS.

INFORMATION REQUIRED TO DESIGN AN ITS

A traditional ITS is divided into four distinct modules depicted in Fig. 2.1: *expert model, student model, instructional model* (which incorporates diagnostic and pedagogical processes), and the *instructional environment*, or user interface (Burton & Brown, 1982; Carr & Goldstein, 1977; Hartley & Sleeman, 1973; Laudsch, 1975; Wenger, 1987).

---

[1]Our purpose here is not to pit experimental design against quasi-experimental design to claim that one is "better" than the other. Rather, the purpose in this chapter is to show that there are alternate ways to design empirical studies, some of which may be better suited to the particular data collection needs and context of the situation than others.

In the subsections that follow, we briefly describe each of these components, including the types of information ITS designers should try to acquire for the specific modules.

Expert Model

The *expert model* contains a representation of the knowledge to be communicated and serves as the standard against which student performance is evaluated. In most cases, the expert model is not only a description of the various concepts and skills that students need to acquire but an actual computational model that can perform tasks in the domain and, in some cases, articulate the rationale behind specific expert actions. It is important to keep in mind that in addition to this knowledge, the ITS designer needs to investigate the problem-solving strategies that are employed by experts in a domain. To build the expert model, ITS designers should secure data that informs them about post- and prerequisite knowledge and relative degrees of difficulty.

Student Model

No intelligent remediation can take place without a certain understanding of the recipient. Therefore, just as an ITS contains an explicit representation of the expert domain knowledge, it must also contain a representation of the student's knowledge in the form of a *student model*. Through the student's interaction with the system, unobservable aspects of the student's behavior are inferred to produce an *interpretation* of his or her actions and to reconstruct the *knowledge* that has given rise to these actions. This information supports detailed pedagogical decisions in guiding the student's problem solving and in organizing his or her learning experiences.

ITSs are knowledge intensive; they are built around practice and feedback to promote *knowledgeable learning* rather than rote learning. Knowledgeable learning attempts to identify missing knowledge and to explicitly remediate misconceptions. One important category of information needed for an effective ITS is an *error taxonomy*. An error taxonomy can be thought of as a catalog of known conceptual, procedural, and strategic misconceptions (i.e., bugs) that is used in diagnosing errors related to procedural sequences, reasoning, and incorrect models of the domain. The classification of errors into an error taxonomy is a beginning attempt at identifying the causes of errors. There is a considerable body of research on the nature of errors and error taxonomies (e.g., Altman, 1967; Chapanis, Garner, & Morgan, 1949; Fleishman & Quaintance, 1984; Rouse & Rouse, 1983; Singleton, 1973). Most error taxonomies can be derived from the type of error (e.g., errors of omission, commission, sequence, slips, and mistakes).

Another important insight about the target population gained through an error taxonomy relates to what concepts students find particularly difficult or easy to learn. This information can be used to focus the tutor on remediating the difficult topics rather than on tutoring easy topics. By focusing on the difficult topics, the designer can determine the class of domain problems that the tutor's expert module should be able to solve.

Instructional Model

Pedagogical decisions are made by references to the student model and to the model of domain knowledge. At the global level, these decisions adapt the sequencing of instruction to the needs of individual students. At the local level, the instructional model determines when an intervention is desirable, whether the student should be interrupted in an activity, and what could and should be said or presented at any given time. This includes guidance in the performance of activities, explanation of phenomena, and remediation.

Wenger (1987) enumerated a number of strategies employed by teachers that facilitate learning (e.g., teaching by analogy, generalization, example, simulation, causal accounts, functional accounts, abstraction, etc.). The designer needs to be aware of the strategies employed by a human tutor to transfer that pedagogical expertise into the tutoring system.

Another important body of information needed for the pedagogical process is the curriculum. An ITS designer needs to know about pre and postrequisite relationships between topics and their relative degrees of difficulty. Having this information is the key to assembling flexible sequences of instruction.

Instructional Environment

The *instructional environment* (or user interface) is the primary communication mechanism between the user and the ITS. As a "communicator," the instructional environment must support several different "channels" of communication. First, the instructional environment must communicate characteristics of the actual task environment pertinent to operations within that environment (e.g., its appearance, its method of interaction, etc.). Second, the instructional environment must have a means of communicating instructional intent and assistance to the user with minimal distraction to the tasks at hand (i.e., what users are expected to do next in the sequence of instructional activities). Finally, the instructional environment must support users' inspections of the ITS's assessment of their knowledge and skills (and sometimes, the rationale behind those assessments).

An additional issue affecting the instructional environment has to do with identifying what segment of a population will be the target of the ITS. An ITS for a complete novice can be quite different than one designed for a more experienced user. Novice users generally require an ITS that is capable of presenting more prerequisite information and providing more procedural and strategic guidance in task performance than an ITS for more experienced users.

Although specific ITSs vary according to how their internal architectures are designed, most researchers have identified these four components—expert model, student model, instructional model, and instructional environment—in much the same way (the interested reader is referred to Wenger, 1987, for a good overview of ITS systems).

In the next section, we discuss some issues surrounding empirical studies of particular import to gathering information for the design of an ITS.

QUASI-EXPERIMENTAL DESIGNS

As Cook and Campbell (1979) observed, efforts to control measurement and treatment variables are exerted to maintain *validity*. That is, the investigator desires to make truth claims about whether the observations made during the study demonstrate cause and effect; the degree to which the cause–effect relationship can be generalized to other populations, situations, and settings; the degree to which the cause and the effect have construct validity (i.e., have a "real" cause and a "real" effect been measured, or have the study variables been confounded?); and the magnitude of the cause–effect relationship.

Cook and Campbell (1979) distinguished among three general classes of quasi-experiments. The first, *nonequivalent group designs*, are those in which responses of both treatment and control groups are measured before and after a treatment manipulation. Nonequivalence in groups can range from single group designs (with no controls—the ultimate in nonequivalence) to multiple group designs in which groups approach equivalence in size, random assignment, and so on. The strength of the test of the quasi-experiment's causal hypotheses is directly related to the degree of nonequivalence employed. Conversely, threats to the validity of nonequivalent group design experiments are inversely proportional to the degree of nonequivalence, with single group designs presenting the most serious threats to validity. As groups become increasingly nonequivalent, it becomes harder to rule out plausible alternate explanations; hence, harder to make reasonable causal inferences.

The second class of quasi-experiments proposed by Cook and Campbell (1979) are called *interrupted time-series designs*. Interrupted time-series designs "are those in which the effects of a treatment are inferred from comparing measures of performance taken at many intervals before a treatment with measures taken at many intervals afterwards" (p. 6).

The third class of quasi-experiments are referred to as *passive observational methods*. Passive observational methods are those in which researchers attempt to "infer causal processes based on observations of concomitancies and sequences as they occur in natural settings, without deliberate manipulation and controls to rule out extraneous causal influences" (Cook & Campbell, 1979, p. 295). In passive observational methods, researchers are interested in discovering whether certain variables covary with others, irrespective of whether any of the variables are manipulated.

An investigator may choose to implement a quasi-experimental design over other types of designs for a variety of reasons such as:

- Constraints of the investigator's operating environment do not permit complete, randomized designs (as discussed earlier).
- The investigator may wish to explicitly avoid the intrusive manipulations and control so as to reduce the effects of observation as much as possible (as in ethnographic studies).
- The environment may not permit a well-controlled design (e.g., in a naturalistic setting such as a classroom).

Before proceeding with a discussion of quasi-experimental designs and their efficacy and appropriateness in gathering information to inform the design of an ITS, we discuss the concept of validity, which is a central concept in all empirical studies.

SOURCES OF VALIDITY

All ITS designers are concerned with ensuring that the knowledge their ITS "tutors" is accurate, the instructional strategies used by their ITS are appropriate, the data the ITS collects and the algorithms the ITS uses to build its student model are sufficiently accurate and sensitive, and the user interface used to emulate the task domain is veridical. At an abstract level, all of these concerns are related to the concept of validity.

Cook and Campbell (1979) used the concept of validity to refer to the truth of a proposition, especially a proposition of causality. These investigators identified four types of validity: internal validity, construct validity, statistical conclusion validity, and external validity. Each of these types of validity and factors (referred to as *threats to validity*) that may jeopardize the interpretability of empirical studies are described next.

Internal Validity and Associated Threats

Internal validity refers to the degree to which we can say that a relationship between two variables is causal or that the absence of a relationship implies the absence of a cause. Thus, when an investigator makes the claim that A causes B, this is a claim of internal validity. This type of validity is central to field research because it allows an investigator to make a claim that introducing a specific treatment *causes* a predictable result. As depicted in Fig. 2.2, internal validity is of particular concern in the gathering of knowledge to inform the design of the expert and instructional models of an ITS.

A large part of what needs to be incorporated into an expert model is the knowledge of operations or behaviors within a domain and how the objects in the domain deal with abnormal situations or emergent phenomena. In poorly formulated domains, for which ITS designers must develop their own expertise, it is critical that they validate the accuracy of this domain expertise. In the case of the instructional module, given the wide range of instructional strategies available, it is important to ascertain which of these instructional strategies affords the most effective learning (hence, is the most valid method to use).

Threats to Internal Validity. Each type of validity has associated *threats* that can jeopardize or nullify the ability of an investigator to detect cause–effect relationships. Threats associated with internal validity that may compromise an empirical study include the following (Cook & Campbell, 1979):

History. This refers to a threat that occurs when an unplanned event takes place between the pretest and the posttest measurements to produce an effect. For example, students learning how to use a tutor may acquire knowledge in the task domain between sessions with the tutor, resulting in performance improvements that may mistakenly be attributed to the tutoring system.

Testing. The number of times a person interacts with a test may directly affect performance on that test as associations between test items and responses are learned.

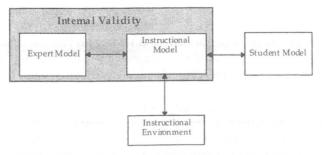

FIG. 2.2. ITS modules most concerned with internal validity.

Instrumentation. If the measuring instrument changes between pretest and posttest, this change may have an effect on measurements, especially if metrics or scales of measurement change with the instrumentation. For example, measurement of domain knowledge/skills in a pretest not administered by an ITS, compared to a posttest administered by a tutoring system, may be compromised if a change in testing and measurement systems exists. This threat can also obtain if the measuring instrument is not sensitive enough to detect further improvements in performance (a *ceiling effect*) or cannot detect performance gradations at low levels of performance (a *floor effect*).

Statistical regression. When participants are assigned to treatment conditions on the basis of pretest scores, this threat may be operating. Regression of pretest–posttest gains scores is to the population mean. As a result, statistical regression may inflate low pretest–posttest gain scores because the performance of these participants was probably depressed due to error, whereas high pretest–posttest gain scores are depressed because the performance of these participants was inflated due to error (participants with pretest–posttest gains scores in the middle of the distribution are not affected as they are already near the mean).

Selection. In quasi-experimental studies, the investigator rarely has the opportunity to randomly assign study participants to treatment conditions. As a result, systematic biases may exist in some conditions that pose a threat to validity. For example, in a study testing the efficacy of an ITS, one session may be composed primarily of people with a high skill level in the domain being taught (e.g., because union rules may require that more experienced employees be given first exposure to new training), and a different session may be composed primarily of novices (because, perhaps, a manager wanted his or her people to get training as early as possible and not be "left out").

Interactions with selection. Any of the aforementioned threats may interact with selection to produce threats to validity. For example, selection–history interactions may occur such that some groups of participants experience unique events between pretest and posttest that other groups do not.

Diffusion of treatments. This occurs when groups of participants communicate with one another such that individuals in one group may learn information that affects their performance. In testing an ITS, for example, some students may inform others about lessons, test answers, and so on, that have not yet been seen, thereby influencing the behavior of the informed participants.

Inequality of treatments. A number of threats arising from perceived or actual inequality between experimental treatment groups and control groups can threaten internal validity. Cook and Campbell (1979) listed at least three such threats: (a) compensatory equalization of treatments, which

occurs when organizational policy requires all groups to receive some perceived benefit and, thereby, disallows a no-treatment control group; (b)
compensatory rivalry among participants receiving unequal treatments may
occur when participants in a control group (who may view themselves as
"underdogs") strive to do as well as those in treatment groups, resulting in
overcompensation. This may occur in IT studies if a vested group, for
example, classroom instructors, views itself as "replaceable" by the ITS. In
these cases, such instructors may overcompensate to demonstrate that they
are "better than" the ITS; (c) resentful demoralization of participants receiving less desirable treatments may occur when the perceived benefits of
being in a treatment group are greater than being in another treatment
group or in a control group. In these cases, the participants in the less
desirable groups may retaliate by depressing their performance.

Construct Validity and Associated Threats

As depicted in Fig. 2.3, a second type of validity, of concern to the expert,
instructional, and student models, is that of *construct validity*. Construct
validity refers to the possibility that there may be more than one cause–effect
explanation for some observed result. For example, a well-known construct
validity concern is the placebo effect. In early medical experiments on drug
effectiveness, the therapeutic effect of the doctor's concern was confounded
with the chemical action of the drug. In other words, researchers were
unable to discern whether the drug (cause 1) produced positive responses
in patients (result) or whether the doctor's concern (cause 2) produced
positive results. To circumvent this problem and produce a result with
construct validity, the notions of double-blind experiments and placebo
control groups were introduced.

Construct validity is of particular concern in the early stages of an effort
when attempting to understand some domain's cause–effect constructs. In
addition, construct validity is of concern in determining the cause of some
misconception in a domain (construct validity of cause) and the most effec-

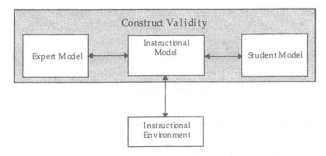

FIG. 2.3. ITS models most concerned with construct validity.

tive way of remediating that misconception to produce optimum learning (construct validity of effect).

Threats to Construct Validity. This class of threats interferes with the ITS investigator's ability to detect whether the tested instructional approach is effective. These threats include:

Inadequate preoperational explication of cause and effect constructs. A clear understanding of the nature of the cause and effect constructs is necessary to determine whether causal relationships exist between these constructs. For ITS studies, it must be clearly understood what it means to acquire knowledge and skills in a domain (effects) and how pedagogical strategies affect such learning (causes).

Mono-operation and monomethod biases. Mono-operation bias occurs when the investigator has only one construct to represent cause and only one measurement for effect. Monomethod bias occurs when all treatments are applied in the same manner and/or all participant responses are measured in the same manner.

Hypothesis guessing. When participants know they are in an empirical study, they may be able to guess the hypothesis or hypotheses being tested.

Evaluation apprehension. Cook and Campbell (1979) cited evidence that some people may be anxious about their skills being evaluated. Such people may feel a strong need to appear intelligent, which may affect their performance.

Experimenter expectancies. This threat refers to the potential for investigators to exert (unknown to themselves) biases over participants' performance due to their expectations about cause–effect relationships. In ITS studies, for example, the investigator may anticipate that a tutoring system will improve student performance relative to a no-tutor condition. This may subtly influence the manner in which the investigator delivers treatments such that the tutor group receives slightly more attention from the investigator than does the no-tutor group, thereby inflating the tutor group's performance.

Confounding constructs and levels of constructs. This confounding occurs when the investigator concludes that A does not affect B when, in fact, some unmeasured level of A does affect B.

Statistical Conclusion Validity and Associated Threats

Statistical conclusion validity is used to determine the probability that a cause–effect relationship exists. Cook and Campbell (1979) listed three decisions for determining statistical conclusion validity: (a) if the study is sensitive enough to detect covariation of cause and effect; (b) if the requisite

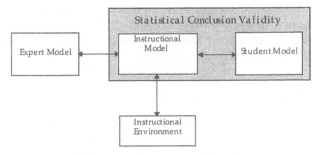

FIG. 2.4. ITS models most concerned with statistical conclusion validity.

sensitivity is present, is there evidence for covariation; and (c) if there is evidence of covariation, how strong is the covariation?

The first issue is one of statistical power. The greater the statistical power, the greater the probability of detecting a cause–effect relationship, if one exists. The second issue has to do with the use of an appropriate statistical test to determine if a cause–effect relationship exists. Researchers speak of the probability of a false alarm in claims that effects exist (the p value) and may use statistical tests to make such determinations of significance. As depicted in Fig. 2.4, statistical conclusion validity is most relevant to the development of the student model and instructional model's diagnostic processes.

As stated previously, student modeling involves inference of student knowledge from observable behavior so that instructional strategies and interventions may be appropriately applied during training. Behavior observable to an ITS includes the performance of task actions at appropriate times and in appropriate sequences and indications of knowledge of concepts and/or the expected behavior of objects within a domain. The questions of whether an ITS is modeling the appropriate actions and accurately assessing student knowledge states are best answered in empirical investigations focused on establishing statistical conclusion validity.

Threats to Statistical Conclusion Validity. Statistical conclusion validity refers to issues regarding error in measurement and the correct use of statistical tests. This source of validity is important to the ITS investigator as he or she strives to measure the efficacy of an ITS, particularly with respect to the system's student model. As discussed previously, the student model is used in ITSs to describe the student's knowledge at, ideally, any given point in an interaction with the ITS. Based on the student model, the ITS may perform such tasks as presenting appropriate feedback or making curriculum selections for the student. An empirical study with low statistical validity will make it difficult for the investigator to determine, from the collected data, how well an ITS or a specific pedagogical approach works.

Threats to statistical validity involve low statistical power (i.e., low ability to detect existing cause–effect relationships), violation of assumptions of statistical tests, inflated error rates resulting from repeated analysis of the same data sets (often referred to as "fishing" for results), and low reliability of measures (i.e., measures that cannot be repeated).

External Validity and Associated Threats

External validity refers to the ability to generalize the findings of a study to different populations, situations, settings, and times. To assure external validity, investigators must identity a target population of participants, settings, and times, and methodically draw samples that represent these target populations. However, samples are often drawn because they are intuitively similar to the target population. Difficulties can arise when samples are drawn from populations that poorly represent the target population on variables of interest (as might happen, for example, when sophomore college students are sampled to represent experienced corporate employees in a test of training methods).

For an ITS study, it is sometimes possible to draw samples from the actual organization for which the tutor is being developed. In such cases, threats to external validity are reduced compared to situations in which populations outside of the receiving organization must be sampled as experimental surrogates for the target population.

Threats to External Validity. Our use of the term "external validity" for ITS studies is somewhat broader than that of Cook and Campbell (1979). Those researchers referred to the ability to generalize results from an empirical study to other situations, settings, participants, and situations.

We refer to external validity in another important sense for ITSs. The degree to which an ITS's training environment provides a faithful simulation of a task domain (e.g., the real-world job an ITS trains students to perform) determines the ITS's ability to teach. The greater the correspondence of the simulation to the task environment, the greater the probability that the ITS will teach effectively. In this sense, external validity refers to the ability to make claims about the generalizability of ITS training effects to performance in the actual task domain.

Interaction of setting and treatment. To what extent can results obtained in one setting, such as a university classroom, be generalized to a different setting, such as a corporate classroom? Can the selected participants in an ITS study be considered representative of the entire user population, or were they selected because they "stood out" for some reason (e.g., perhaps they were low performers who are considered prime candidates for additional training, or they were highly motivated individuals who wanted to experience new technology as soon as possible)? When these conditions obtain, it is difficult to generalize to a different user population.

Summary

In well-formulated domains, ITS designers are less concerned with empirical evidence of validity for the expert and instructional modules, opting instead to rely on the expertise of their subject-matter experts and instructional designers. The general principle in well-formulated domains is to "take for granted" the expertise of subject-matter experts and to accept that expertise as subjective evidence of validity. However, even in well-formulated domains, ITS designers should be concerned with obtaining empirical evidence of the statistical conclusion validity of the student modeling and diagnostic processes and the external validity of the ITS in general.

In poorly formulated domains, ITS designers should be concerned with obtaining empirical evidence of the validity of all of their ITS modules. As stated previously, in poorly formulated domains, sufficiently mature expertise does not exist, and, hence, cannot be sufficiently articulated by domain resources (i.e., subject-matter experts and instructors for that domain). However, it is easy to fall into the trap of failing to recognize the domain as poorly formulated, resulting in the use of naturalistic observation methods (e.g., traditional knowledge-acquisition methods) when empirically based methods are more appropriate. The end result is ITSs that are of questionable value because they contain uninformed views of the task domain and of instruction within the task domain.

Regardless of how well formulated a domain may be, it is important for the ITS designer to bear in mind that each type of validity has associated threats that must be addressed in any empirical study (we consider specific methods for addressing some of these threats in the conclusion to this chapter).

In the sections that follow, we provide examples of two quasi-experimental studies used to gather design information for an ITS. The first is a study of learning how to do spreadsheet programming in which a combination of passive observational methods and experimental designs were employed.

The second is a quasi-experimental study of how students learn to develop object-oriented models. In this study, students were given specific tasks to perform in the classroom, and these task sequences were manipulated as independent variables between classes. This study represents an example of the use of a nonequivalent groups design.

THE EXCEL TUTOR

The Spreadsheet Tutor was developed as a part of a doctoral dissertation project by one of the authors (Shahidi, 1993). The purpose of this research was to evaluate the effectiveness of student modeling in ITSs. Developing a student modeling component for a tutoring system is very labor intensive.

This study investigated whether the labor of developing student models for an ITS is worth the cost. Two types of ITS were compared—one with student modeling and one without. The two types of ITS were also compared with a traditional source of information available to the student in a software environment—online help—to provide a comparison condition for evaluation.

The Spreadsheet Error Taxonomy

In ITSs, a catalog of known conceptual and procedural bugs is often used to diagnose errors related to procedural sequences, errors in reasoning, and underlying misconceptions reflecting incorrect models of the domain. One of the most important goals of an empirical study is to gather data about likely errors and misconceptions that are committed by students in task domain, so that a *bug library* can be designed for the student model (VanLehn, 1988). Although theoretically limited in scope, this enumerative approach has the practical advantage that the knowledge about the likelihood of errors that is incorporated in the system can be derived empirically or obtained from expert teachers.

An error is a discrepancy between a user's actual performance and the desired performance. Remedial action is usually taken based on a hypothesis of a predisposing factor. It is not sufficient to simply say that an error was made because the user needs more training. The process of remediation works backwards from an error to a specific predisposing factor that caused the error. The classification of errors into an *error taxonomy* for the domain is a beginning attempt for the determination of the causes of errors in a system.

In addition to using type of error (e.g., errors of omission, commission, sequence, slips, and mistakes) to construct an error taxonomy, other variables can be used to derive taxonomies, as enumerated by Meister (1989):

- Errors associated with specific equipment or facility.
- Errors associated with a stage of a multistage process (e.g., take-off or landing).
- Errors associated with a specific procedure or a step in the procedure.
- Errors associated with a behavioral function (e.g., decision making, tracking).
- Error consequences (e.g., catastrophic effect or no effect at all).

It is important to note that not every tutoring system requires an error taxonomy. Some tutoring systems can generate bugs on the fly (e.g., BUGGY [Brown & Burton, 1978] and WEST [Burton & Brown, 1976]). These systems tend to deal with domains that are rather small and can accommodate a generative theory of bugs.

TABLE 2.1
Error Types and Example Misconceptions for Spreadsheet Coaching System

Type of Error	Example Misconception

Behavioral	Cell entries
	    Editing cell entries
	    Removing a cell's contents
	    Clearing a range of cells
	    Moving text
	    Naming ranges

Intentional Errors in Worksheet Planning	Identifying when to apply each addressing mode
	    Choosing relative vs. absolute vs. mixed
	    (hardest concept to teach)
	    Identifying ranges
	    Positioning formula

Slips	Not hitting ENTER key after an entry Making typing mistakes Not noticing which mode you are in Choosing the wrong menu item

As discussed previously, error taxonomies in ITSs are usually based on identification of types of errors students may commit in a particular task domain. For the spreadsheet domain, those types were behavioral errors, intentional errors, and slips. Behavioral errors are mostly syntax errors and lack of knowledge of how to operate the spreadsheet program. Intentional errors are problem specific and show lack of knowledge on the part of the student in applying the right concepts at the right time. Slips are purely careless mistakes that result from lack of attention or concentration.

Expert opinion was used to make a first pass at developing an error taxonomy for spreadsheet programming: During the first phase of the study, a series of interviews was conducted with five Excel instructors at the University of Pittsburgh's Computer Learning Center. During the second phase, students drawn from the university population were observed during a problem-solving session.

The interviews revealed an error taxonomy for Spreadsheet students, fragments of which are shown in Table 2.1 (see Shahidi, 1993, for a complete taxonomy).

Errors committed in spreadsheet basics seem to be all behavioral, whereas errors in spreadsheet planning seem to be all intentional. Of course, this domain has its share of slips, which account for noise in the diagnostic process.

Error Taxonomy Validation Using Passive Observation

Gagne, Briggs, and Wagner (1988) and Golas (1983) identified one-to-one testing as an important step early in development to minimize inappropriate development. In one-to-one testing, observers make detailed observations of how a student interacts with the task environment. Investigators can observe student capabilities, identify inappropriate expectations, and in this case, validate the error taxonomy.

Letters of invitation were sent to 50 graduates of the Computing Learning Center Excel® course who had taken the course within the past 6 months. Four people accepted the invitation to participate. In the preliminary questionnaire, the participants were asked to rate their frequency of using Excel by choosing among very little, often, frequently, and very heavily. One person used Spreadsheets frequently, two often, and one very little.

The students took approximately 45 minutes to solve an Excel problem. They were encouraged to use online help. If they could not find an answer through online help, they were allowed to ask questions from the investigator. This allowed us to observe the limits of online help. The observer, therefore, also had the role of simulating the ITS.

Notes were taken about each user's errors. All participants made numerous mistakes, and none of them had a completely solved spreadsheet by the time they thought they were done.

Results. The Computing Learning Center course (which all participants had taken before participating in our study) had effectively taught the students how to enter and remove information from spreadsheet cells. Even the student who seldom used Spreadsheets did not make errors of that type. As expected, most of the errors concerned the use of formulas. Although the participants made many temporary errors that covered more of the taxonomy, they were often able to correct the errors after seeing the effects on the screen. As most of the errors had to do with formula entry, it was decided to have the ITS concentrate mainly on this topic.

Testing the Efficacy of Student Modeling

A research interest in constructing the Spreadsheet Tutor was to determine the extent to which student modeling assists students in learning spreadsheet programming. An experimental study was conducted to investigate this issue.

Three groups of participants, with 30 in each group, were involved in the study. The treatments in each group determined what type of coaching the participants received while solving spreadsheet programming problems, specifically: (a) coaching with student modeling: The system made context-sensitive generalizations and distinctions by referencing students' past actions and exploiting relationships among concepts in the domain; (b) coaching without student modeling: The system provided surface feedback regarding the correctness of procedures, but the learner was expected to identify and carry out correct procedures; and (c) online help only: It was up to the learner to both determine the correctness of his or her procedures and to identify correct procedures to remedy errors. Each group received 3 hours of classroom instruction on spreadsheet programming before being asked to solve a set of training problems. Finally, participants in each group took a posttest to assess their knowledge.

Results. The participants who interacted with the ITS (group a) showed the highest level of learning. Those who interacted with the rudimentary ITS (group b) showed performance equivalent to those who did not interact with any type of ITS (group c). Further analysis of student errors indicated that what seemed to matter most was the ITS's anaphoric reference to students' prior actions and explicit feedback on how to transition from a current knowledge state to the next state.

This use of both quasi-experimental and experimental methods illustrates how the ITS designer can customize the design of an empirical study. In validating error taxonomies, passive observational studies or other types of quasi-experimental studies are appropriate in enumerating and classifying student errors. When it comes to validating student models and their efficacy, empirical methods that emphasize construct validity and statistical conclusion validity are appropriate.

In the next case study, we examine a quasi-experimental approach to collecting data in a corporate setting—an environment in which constraints on time and personnel resources required for an experimental study of the sort conducted for the Spreadsheet Tutor are not typically available.

AN INTELLIGENT TUTORING SYSTEM FOR OBJECT-ORIENTED MODELING

The recent move toward business process re-engineering has caused many organizations to look at object-oriented software development as a way to re-engineer their software infrastructure (Jacobson, Christerson, Jonsson, & Overgaard, 1992; Keyes, 1992). As a result, object-oriented development is viewed at many large corporations to be a key success factor in re-engineering legacy systems (software infrastructure that may be 20 or 30 years old) into new, more efficient systems.

Because of the shortage of highly skilled, object-oriented practitioners available for hire, however, the only avenue open to many organizations is to train existing software developers in the object-oriented paradigm.[2]

Training a large workforce of object-oriented practitioners is an almost overwhelming challenge. Because object-oriented skill acquisition is not well understood from a psychological/human learning perspective, it is difficult for any organization to select, much less evaluate, training vendors. What is well understood about learning object-oriented skills, however, is that much more training and practice is typically needed than an organization can afford to provide its employees. Many software development managers are, therefore, caught in the following quandary:

1. Object-oriented software development is viewed as a success factor for reengineering.
2. Object-oriented practitioners are difficult to find in numbers sufficient for a large organization; therefore training is required.
3. Training is expensive and cannot be delivered fast enough to solve the immediate need for expertise in object-oriented modeling and programming.

In the following sections, we outline our approach to addressing these difficulties.

The SLOOP System

We have begun to address the need for an effective way to teach object-oriented concepts by investigating how learning occurs in the corporate classroom. This work, as with most industry-initiated research, has a very pragmatic aspect. In this case, the challenge is to build an ITS that supports students as they practice newly acquired skills in object-oriented requirements modeling, analysis modeling, and design modeling.[3] We refer to this ITS as SLOOP (System for Learning the Object-Oriented Paradigm).

SLOOP provides feedback to students as they work through sequences of problem-solving activities to create object models in the requirements, analysis, and design phases (the nature of these models is described in more detail in the next section). As students progress through these activity

---

[2]The problems are obviously the greatest for large organizations (e.g., on the order of many hundreds or thousands of employees). For smaller organizations, hiring small numbers of object-oriented software developers may be a realistic alternative to training.

[3]The development of the Intelligent Coaching System is being performed in conjunction with the University of Pittsburgh, specifically: Edward Hughes, graduate student, Department of Computer Science; Professor David Hurley, Department of Computer Science; Professor Alan Lesgold, Associate Director, Learning Research & Development Center.

1. A consortium consists of banks working together.
2. The consortium owns a number of automated teller machines (ATMs).
3. An ATM is for entering ATM transactions.
4. ATM transactions are for updating bank accounts.
5. The consortium owns a central computer.

FIG. 2.5. Sample requirements statements.

sequences, they practice (and most likely, relearn) rudimentary modeling skills to which they have just been exposed in a classroom setting. Additionally, students practice a *process* for developing object-oriented models.

Modeling in SLOOP. Students begin a problem-solving session by viewing a collection of requirements statements (see Fig. 2.5). These requirements statements, which represent a customer's requirements for the system that is to be modeled, have been carefully constructed so that each statement expresses a single modeling concept such as a relationship, an attribute, an operation, and so on.[4]

From these requirements statements, students must identify noun phrases, as these provide clues to the various objects, attributes, and operations that will be in the resulting model (this is similar to the approach taken by Wirfs-Brock, Wilkerson, & Wiener, 1990, among others). Students are then asked to group noun phrases into specific, defined categories such as irrelevant noun phrases, qualifiers that do represent objects, noun phrases that name abstract concepts, class objects, actors, attributes, and so forth. After all noun phrases have been identified and categorized, students are ready to begin building object models.

There are a variety of object-oriented methods from which to choose. This potentially puts us in a difficult situation for the SLOOP ITS in that any specific methodology we select to coach almost certainly does not meet the needs for all of our end users (more detailed descriptions of our end-user populations are provided later; for now, it is sufficient to assume that anyone who needs to learn object-oriented techniques in a corporate setting falls into the target audience). Because some end users prefer OMT (Rumbaugh, Blaha, Premerlani, Eddy, & Lorenson, 1991), whereas others prefer OOSE (Jacobson et al., 1992) or OOD (Booch, 1994), the approach we have adopted for the ITS is to use *model-based development.*

In model-based development, a set of abstract models is used to produce more concrete specific models. In this case, we start with an informal model

---

[4]Most object-oriented methodologies start with some form of statement of customer needs, though to our knowledge, SLOOP is more strict in how these requirements statements are represented. We realize that learning how to construct requirements statements is a skill in itself, antecedent to developing object-oriented models, but is beyond the scope of teaching object-oriented modeling. We are currently investigating ways to teach this skill as well.

of user requirements and proceed through progressively more formal models of analysis modeling and design modeling, ending with a specific model of implementation represented in a programming language.[5] Because this approach allows us to narrow the "semantic gap" between the various object-oriented methods, it provides the flexibility required to coach students in any of the established object-oriented methods. This also affords us the interesting option of teaching a "generic" process for object-oriented modeling that combines the most interesting aspects of the various object-oriented methods. These abstract models provide the foundation for development of modeling processes for requirements modeling, analysis modeling, and design modeling in the SLOOP system.

A Quasi-Experiment in the Classroom

The domain of object-oriented modeling falls into the category of poorly formulated domains, as discussed in the introduction to this chapter. This usually means that there are no formally established procedures and expertise for the task domain or for training in the domain. In the case of object-oriented modeling, it is more accurate to say that there is a plethora of modeling methodologies and teaching methods but that there is no clearly defined paradigmatic approach to the task domain. As a result, the SLOOP development team[6] found itself in the position of needing to quickly and empirically identify effective teaching strategies that could be incorporated into SLOOP. We set about this task by collecting data from students learning object-oriented skills in the classroom.

We had a number of goals for gathering data in the classroom. First and foremost, we wished to gather information about the students from the populations of US WEST employees who will eventually be users of SLOOP. We wanted to know what knowledge students already had about software design in general and object-oriented techniques in particular. Second, we wanted information about both the material being taught and strategies for teaching it. That is, we wanted to gain insight about which object-oriented concepts are difficult to understand and prone to misconceptions while observing the effect of teaching strategies for overcoming these problems. Finally, we wanted to determine how much classroom instruction is required before students can begin to effectively use an ITS (we were targeting students who had completed an introduction to object-oriented analysis and design as the target user population). As a practical matter then, it was

---

[5]The material in this section was developed by David Hurley.

[6]In addition to the aforementioned group at the University of Pittsburgh, the SLOOP development team included the following US WEST Technologies members at the time this chapter was written: Charles Bloom, Patricia Haley, Srdjan Kovacevic, Anne McClard, Michelle Neves, Anoosh Shahidi, and Scott Wolff. Charles Hymes and Robert Rehder contributed to the project as summer interns.

important for us to determine how extensive such a course must be before students possess the minimum understanding of object-oriented concepts needed to make use of an intelligent coaching environment.

In terms of Cook and Campbell's (1979) distinction among types of quasi-experiments, we conducted several one-group, pretest–posttest designs, that is, we considered the students in each training session we conducted to be a group. By presenting students with paper-and-pencil problem-solving activities that mirrored those of the SLOOP system, we were able to collect data about the types and categories of student errors (allowing us to create and validate a SLOOP error taxonomy). In addition, by varying the sequences of problem-solving activities among classroom groups, we were able to test assumptions about how best to teach the material.

Collecting Data in the Classroom

The course we designed was divided into three sections. First, we presented an overview of the so-called "software engineering crisis" (e.g., Booch, 1994) for which object-oriented design techniques are intended to be at least a partial solution. Second, we presented object-oriented modeling as a general design technique divorced from a particular software design methodology. Our intent in this section was to introduce students to elementary object-oriented modeling constructs such as classes, attributes, relationships, and so on. Third, we taught an object-oriented software development based on the SLOOP abstract models. In this final section of the class, students were exposed to the concepts of requirements models, analysis models, and so forth, and rules for generating one model from another. Our class was originally designed as a one-day session, but our experiences with this compressed agenda, along with our desire to collect additional data from the students, led us to eventually extend the class to 2 days. This class was taught five times (with additional sessions being planned at the time this chapter was written).

Data from our class came from a variety of sources. First, we used an object-oriented modeling example that was constantly expanded during the course of the lecture. Because the SLOOP system is being designed for students who have had an initial introduction to object-oriented analysis and design, we wanted students to practice solving at least one problem in class before collecting hard data on a second problem. This problem, which involved modeling a hazardous waste storage facility (the problem was adapted from Coleman, Arnold, Bodoff, Dollin, Gilchrist, Hayes, & Jeremaes, 1994, and cast in a form consistent with the way problems are presented in SLOOP) was solved interactively, with students given time to complete parts of the problem, followed by instructor-led discussion of their proposed solutions. The only data collected during these discussions was a list of issues raised by students as they talked about the problem with the instructor and other students.

Once the lecture phase of the course was completed and students had completed the in-class hazardous waste storage facility problem, they were presented with a second problem, the now-classic automated teller machine (ATM) problem. Their solutions to this problem were our second source of data from the classroom. However, in contrast to the interactive style with which students solved the first problem, our approach was to present the ATM problem using a semiformal simulation of the pedagogical method we had designed (i.e., we did not carefully counterbalance experimental conditions or seek to control error variance) but not yet implemented for the SLOOP system. In other words, our goal was to simulate the SLOOP pedagogy as much as possible with paper and pencil.

To simulate the environment of an interactive ITS as closely as possible, students were told that they were not taking a test, that they were free to raise any questions they wished during problem-solving activities, and that they could have as much time as they needed to complete each activity.

Students' answers were recorded on sheets that were collected by instructors before proceeding to the next activity. Correct answers to questions were typically provided individually to the student and took the form of hints and suggestions as might be offered by an ITS. Feedback was also provided by asking the entire class if there were questions about the problem-solving activity that had just been completed. When questions were raised on this level, the instructors held a class discussion before proceeding to the next activity. The questions raised by students, both individually and during class discussions, were recorded by the instructors.

Finally, we asked students to complete pre- and postclass questionnaires. This provided us with information about each student's programming experience and background as well as their impressions of the course as a learning experience. The questions related to students' prior knowledge of computer programming were used as a surrogate for a pretest. These "pretest" questions were directly related to the variables of interest, that is, computer programming skills, and were used to look for correlations between prior experience with procedural languages and performance on the object-oriented problems solved in class. In addition, they provided us with a means of grouping students into equivalence groups based on prior experience.

TESTS OF THE SLOOP PEDAGOGICAL APPROACH

To understand the efficacy of the problem-solving approach we had designed for SLOOP, we collected data from students as they created requirements models for the automated teller machine (ATM) problem. Before our training sessions, we speculated about what was difficult for students in requirements modeling. This list included such items as:

- Identifying actors, class objects, attributes, operations, and so on, from a list of customer requirements statements.
- Distinguishing among actors, class objects, and attributes.
- Specifying cardinality of class objects and actors.
- Identifying structuring relationships among class objects/actors.
- Identifying the relationship type in a structuring relationship.
- Identifying interactions in dynamics modeling.
- Distinguishing between interactions and operations.
- Identifying interaction types correctly.

Although this is a partial list, it illustrates the type of data we intended to collect during classroom problem-solving sessions. We taught a total of five 1- and 2-day sessions and collected data from the last two 2-day sessions. This provided us with quantitative data on 14 students. Teaching multiple sessions allowed us to make changes to the problem-solving activities presented in the first class that could then be tested with the second class. Although our data are preliminary and based on small sample sizes, we have begun to confirm several of our hypotheses:[7]

- Students who were required to identify and remove all noun phrases that represent nonattributes (e.g., qualifiers, irrelevant noun phrases, abstract concepts, etc.) before being asked to identify attributes, correctly identified more attributes than students who were allowed to select attributes from the entire list of noun phrases (86% vs. 65%). Even though this result may appear obvious by virtue of the fact that students could be expected to make more "hits" when many of the nontarget items are removed, it is an important observation because students had previously complained about the somewhat tedious task of categorizing all noun phrases before being allowed to search for such target items as attributes, class objects, and so on. Additionally, we did not always find that removing nontarget items made a difference in how well students performed in other tasks, as shown in the next item.

- Students who were asked to identify object classes and actors from a list that contained *only* these elements performed as well as students who were asked to identify object classes and actors from a list that contained all noun phrases from the requirements statements. In each case, students correctly identified about 60% of the object classes and 60% of the actors.

---

[7]Thanks to Charles Hymes for conducting this data analysis.

From class discussions and our data, it was also apparent that students had the most difficult time distinguishing between actors and object classes.

• In the first class, students were asked to draw structural relationships by examining each requirements statement and constructing a binary (i.e., single target, single source) structural relationship, if one existed in the requirements statement. Students were given answer sheets with a drawing area beneath each requirements statement in which they were asked to sketch any existing relationship (requirements statements were carefully constructed so that if a statement did contain a relationship, it contained a single binary relationship). In the second class, students were not constrained to draw binary structural relationships but were only told to examine the requirements statements for clues as to what structural relationships exist. Students in this class were provided with answer sheets that contained graphical representations of class objects and/or actors, with enough white space to draw an arbitrary number of relationship links among the elements.

Students who constructed sentence-by-sentence binary relationships correctly represented fewer relationships than did students who were not so constrained (62% vs. 92%, respectively). On the other hand, students who were constrained to construct binary relationships missed slightly fewer relations than did the nonconstrained students (17% vs. 29%, respectively). Thus, it would appear that allowing students to construct graphical models that simultaneously integrate all relationships in the requirements model is a preferred approach to constructing several binary relationship models. We have not yet explored the option of asking students to construct integrated models by reading through the requirements statements one at a time. Presumably, this would reduce the "miss" rate for identification of relationships.

• From classroom discussions and our data, identifying structural relationship types and interaction types was, as anticipated, a difficulty for students. Based on these findings, we decided to limit the types of relationships and interactions students must identify in the early stages of training/tutoring (by focusing on the most important structural relationships, e.g., whole–part).

• Students in each class were asked to justify most of their modeling actions (e.g., creating a class object/actor, relationship, or interaction) with a reference to a requirements statement that justified that operation. This provides backward traceability in the sense that the existence of each object, relationship, attribute, and so forth, in the requirements model can be shown to be derived from at least one (and perhaps more than one) requirements statement. Our data suggest that students who did justify their operations performed better overall than students who did not. Thus, what is a good software development practice may also be an effective training technique.

In addition to fine-tuning the content of the pedagogical activities, our quasi-experiments in the classroom allowed us to test, verify, and extend the SLOOP error taxonomy. This taxonomy is partitioned according to the problem-solving activities students will follow when building object models in SLOOP.

Within each modeling activity, two types of student errors were identified: *misconceptions* and *missing conceptions*. The former, misconceptions, refers to the misapplication of a concept or concepts in a particular problem-solving situation. In this case, the student appears to understand the concept to some degree, but has overgeneralized or undergeneralized it in a problem-solving context. For missing conceptions, on the other hand, the student does not demonstrate any understanding or knowledge of the concept.

For example, a misconception occurs when a student correctly identifies an operation but indicates an incorrect requirements statement to justify the existence of the operation. If the student fails entirely to identify a requirements statement that does in fact contain an operation, a missing conception occurs.

Further quasi-experimental investigations in the classroom and with individual students will allow us to refine this taxonomy and will continue to provide useful information about skill acquisition for object-oriented modeling.

DISCUSSION

When threats to validity obtain in an empirical study, they seriously undermine the investigator's ability to detect causal relationships. In experimental studies, many of these difficulties are addressed by random assignment of participants to treatment groups and control of known factors that introduce error into measurement (as discussed in the introduction to this chapter). In this final section, we address the important issues of how to identify and address threats to the validity of empirical studies, with specific reference to the domain of ITSs.

Addressing Threats to Internal Validity

In ITS studies, which are often quasi-experimental in nature, the threats of history, testing, and instrumentation must be clearly identified and addressed by the investigator because randomized assignment of participants to treatment conditions is typically not possible. This leaves the threats of selection, diffusion of treatments, and issues related to inequality of treatments.

Selection may be a problem, particularly in corporate settings, because the investigator may not be able to select the study participants directly but may need to rely on self-selection or selection by others (e.g., management). In this case, the experimenter may be able to administer pretests and questionnaires to perform a *post hoc* evaluation of systematic biases that may exist in the treatment and control groups.

Issues related to perceived inequality of treatments cannot be addressed by random assignment of participants and can be particularly troubling for the ITS investigator, as ITS technology may be typically perceived as inherently more interesting and exciting to participants than traditional (usually classroom) training. In addition, corporate/union policies may make it difficult to conduct studies with inequality among treatment groups (assuming that sufficient numbers of study participants exist to permit the design of multiple treatment groups) or may require treatment groups that interfere with the study (e.g., requirements that senior union people have first access to the new technology, even if those participants are not in need of additional training). These situations must be addressed according to organizational policy. One approach in such situations is to conduct the necessary number of studies to meet policy requirements but to ignore data from irrelevant groups (e.g., expert skill groups).

Baseline data for comparison to experimental data may often be collected before the ITS study begins, such as the approach we adopted for SLOOP. In this case, we began collecting data about how students acquire object-oriented skills prior to building the tutoring system. This permits the investigator to avoid the issue of who gets the tutoring system, and who does not, and provides baseline data for later comparison of tutor versus no-tutor conditions. Of course, in cases such as this, the baseline data must be collected from the same student population (though not the same individuals) that will receive training on the tutoring system.

In situations where resentment is expected from participants who are either in less desirable treatment groups or who feel threatened by the ITS technology, education about the intended nature of the system being tested can be useful (e.g., is this system really intended to replace instructors, or, similar to many ITSs, is it intended to be an augmentation to classroom instruction—i.e., a tool for the instructor to use?). Of course, enrollment of the relevant stakeholders (i.e., management, student populations, instructors, etc.) at the beginning of the ITS project can often reduce such feelings of inequality (see, for example, the chapter *Introducing Advanced Technology Applications Into Your Organization* in this volume for a discussion of establishing stakeholder buy-in).

Addressing Threats to Construct Validity

In ITS studies, it must be clearly understood what it means to acquire knowledge and skills in a domain (effects) and how pedagogical strategies affect such learning (causes) to avoid threats resulting from inadequate preoperational explication of cause and effect. In well-understood domains, much of this work may already have been done for the investigator. For other domains, such as object-oriented modeling, the investigator may need to explicate cause-and-effect constructs as clearly as possible him or herself. Additionally, the investigator can perform knowledge-engineering activities to gather as much information about what is known about teaching in the domain (e.g., through expert interviews, field observation, literature reviews, etc.). Other threats to construct validity can be addressed as follows:

To *address mono-operation and monomethod biases*, Cook and Campbell (1979) suggested varying treatments effects and measurement contexts when possible. For example, in an ITS study, it may be possible to vary the content of feedback to students (e.g., in terms of the amount of information provided) and to vary the context in which feedback is provided (e.g., tutor vs. classroom instructor).

With respect to *hypothesis guessing*, one way we have avoided this threat is to collect data in a natural setting—that is, as part of a classroom training session. We tell participants that wish to learn from them how they solve problems so that we can design a tutor to assist other students with these activities, and we explicitly avoid any terms that suggest experimentation/quasi-experimentation (e.g., mentioning variation of treatments across training sessions).

As most ITS studies require testing or evaluation of participants' responses to some extent, the threat of *evaluation apprehension* must be taken seriously. We have addressed this issue in the SLOOP quasi-experiments by bluntly stating that students are not being tested and that their responses will be coded for confidentiality—no one will ever know which person completed which response sheets, and so on.

In ITS studies, the investigator may anticipate that a tutoring system will improve student performance relative to a no-tutor condition (*experimenter expectancies*). This may subtly influence how the investigator delivers treatments, such that the tutor group receives more attention from the investigator than does the no-tutor group, thereby inflating the tutor group's performance. This problem is difficult to overcome in ITS studies because of the high visibility of the treatment conditions (e.g., tutoring system vs. no-tutoring system), such that investigators who do not possess preconceived notions about the treatment effects may be impossible to employ. Thus, it is up to the investigator to be wary of such biases and to strive to avoid them.

In ITS studies, *confounding constructs and levels of constructs* can cause the investigator to draw incorrect conclusions about the efficacy of an ITS. For example, the investigator may notice that students do not learn from feedback following an error. If, however, the investigator were to manipulate some relevant variable(s) with respect to feedback, such as anaphoric reference to previous errors and solutions, it may be found that specific levels of anaphoric reference in feedback do result in performance improvements.

Addressing Threats to Statistical Validity

Perhaps the best way to avoid these threats is the proper application of statistical knowledge. If the ITS investigator is not comfortable with statistical analyses, he or she should work with a consultant in this area.

Addressing Threats to External Validity

Cook and Campbell (1979) identified three approaches to addressing threats to external validity, all based on methods for selecting study participants and assigning them to treatment conditions. These include: (a) random sampling of participants and random assignment of participants to treatment conditions, (b) deliberately sampling in a nonrandom manner to obtain "quotas" of participants that address the major groups of interest, and (c) impressionistic sampling in which instances of groups to which the investigator wishes to generalize are selected for participation. These three methods can be used in combination with one another to take advantage of the resources available to the investigator.

CONCLUSIONS

We have demonstrated the following concepts in this chapter:

- That empirical studies are a valuable means of gathering information to inform the design and test the efficacy of ITSs.
- There are a variety of empirical study designs available to the ITS researcher.
- That quasi-experimental studies are a useful way to gather information in corporate settings, where resources (especially time and availability of study participants) are in short supply and randomized experimental methods are often not tolerated or practical.
- That there is a close mapping between types of validity in empirical studies and information requirements for designing ITSs.

Through application of empirical methods, whether randomized–experimental, passive–observational, or quasi-experimental, combined with due diligence in addressing threats to validity, the ITS researcher can obtain a wealth of information for the design of an ITS.

REFERENCES

Altman, J. W. (1967). Classification of human error. In W. B. Askren (Ed.), *Symposium on reliability of human performance in work* (pp. 5–16). Report AMRL-TR 67-88, Wright-Patterson AFB, Ohio: Aerospace Medical Research Laboratories.

Brown, J. S., & Burton, R. R. (1978). Diagnostic models for procedural bugs in basic mathematical skills. *Cognitive Science, 2*, 155–191.

Burton, R. R., & Brown, J. S. (1976). A tutoring and student modeling paradigm for gaming environments. In R. Colman & P. Lorton, Jr. (Eds.), *Computer science and education* (pp. 236–246). *ACM SIGCSE Bulletin, 8.*

Burton, R. R., & Brown, J. S. (1982). An investigation of computer coaching for informal learning activities. In D. Sleeman & J. S. Brown (Eds.), *Intelligent tutoring systems* (pp. 79–98). New York: Academic Press.

Booch, G. (1994). *Object-oriented analysis and design* (2nd ed.). Redwood City, CA: Benjamin-Cummings.

Carr, B., & Goldstein, I. P. (1977). *Overlays: A theory of modeling for computer aided instruction.* AI Lab Memo 406 (Logo Memo 40). Cambridge, MA: MIT Press.

Chapanis, A., Garner, W. R., & Morgan, C. T. (1949). *Applied experimental psychology: Human factors in engineering design.* New York: Wiley.

Coleman, D., Arnold, P., Bodoff, S., Dollin, C., Gilchrist, H., Hayes, F., & Jeremaes, P. (1994). *Object-oriented development: The fusion method.* Englewood Cliffs, NJ: Prentice-Hall.

Cook, T. D., & Campbell, D. T. (1979). *Quasi-experimentation: Design and analysis issues for field settings.* Chicago: Rand McNally.

Fleishman, E. A., & Quaintance, M. K. (1984). *Taxonomies of human performance: The description of human tasks.* New York: Academic Press.

Gagne, R. M., Briggs, L. J., & Wagner, W. W. (1988). *Principles of instructional design.* New York: Holt, Rinehart & Winston.

Golas, K. (1983). The formative evaluation of computer-assisted instruction. *Educational Technology, 23*(1), 26–28.

Hartley, J. R., & Sleeman, D. H. (1973). Towards intelligent teaching systems. *International Journal of Man–Machine Studies, 5*, 215–236.

Jacobson, I., Christerson, M., Jonsson, P., & Overgaard, G. (1992). *Object-oriented software engineering: A use case driven approach* (revised fourth printing). Wokingham: Addison-Wesley.

Keyes, J. (1992, June). Code trapped between legacy, object worlds. *Software Magazine*, 39–45.

Laudsch, J. H. (1975). Some thoughts about representing knowledge in instructional systems. *Proceedings of the Fourth International Joint Conference on Artificial Intelligence* (pp. 122–125). Tiblisi, USSR.

Meister, D. (1989, November). The nature of human error. *IEEE Global Telecommunications Conference (GLOBECOM-89)* (pp. 27–30). Dallas, Texas. New York: Institute of Electrical and Electronic Engineers.

Rouse, W. B., & Rouse, W. H. (1983). Analysis and classification of human errors. *IEEE Transactions on Systems and Cybernetics*, SMC-13, 539–549.

Rumbaugh, J., Blaha, M., Premerlani, W., Eddy, F., & Lorenson, W. (1991). *Object-oriented modeling and design.* Englewood Cliffs, NJ: Prentice-Hall.

Shahidi, A. K. (1993). *Evaluation of an approach to intelligent coaching with student modeling.* Unpublished doctoral dissertation, University of Pittsburgh.

Singleton, W. T. (1973). Techniques for determining the causes of error. *Applied Ergonomics*, 126–131.

VanLehn, K. (1988). Student modeling. In M. Polson & J. J. Richardson (Eds.), *Foundations of intelligent tutoring systems* (pp. 55–78). Hillsdale, NJ: Lawrence Erlbaum Associates.

Wenger, E. (1987). *Artificial intelligence and tutoring systems.* Los Altos, CA: Kaufmann.

Wirfs-Brock, R., Wilkerson, B., & Wiener, L. (1990). *Designing object-oriented software.* Englewood Cliffs, NJ: Prentice-Hall.

3

COST–BENEFITS ANALYSIS FOR COMPUTER-BASED TUTORING SYSTEMS

A. Scott Wolff
US WEST Advanced Technologies

In a research and development environment, the decisions to undertake, fund, continue, and cancel projects are constantly reviewed. Under ideal conditions, decisions such as these are always made after carefully evaluating the costs and benefits involved. Many conditions, such as unfamiliarity with cost–benefits analysis, client demands, and budget cycles may force decisions before all the relevant data can be considered.

There may be technical difficulties in conducting cost–benefits analyses: Data may be unavailable or incomplete, all relevant factors influencing cost and benefits may not be understood, it may be difficult to accurately estimate the probability of an event relevant to the decision, and so on. Despite these difficulties, however, it is usually worth the effort to identify the major sources of costs and benefits and to estimate the probabilities of potential outcomes before making important decisions—such as whether to pursue a new technology or to cancel a project. Failure to look at the relevant costs and benefits can lead to decisions that are ill informed and have unexpected outcomes.

Although cost–benefits analysis can reveal advantages and shortcomings of a technology in terms of financial considerations, there are advantages related to adopting a new technology because it provides value as perceived by the customer. These so-called *differentiation advantages*, which include such things as perceived value of new technology for its own sake, advantages in motivating students to use computer-based systems over traditional approaches, and advantages related to introducing new technologies to an organization, translate into services and goods that customers want and are willing to purchase. Thus, there is a danger in focusing on cost considerations exclusively—both cost and differentiation advantages should be considered in evaluating choices, though as Sherman (1990) noted, "It is almost impossible to maximize cost and differentiation simultaneously in a sustainable, profitable manner" (p. 14). To balance these issues, an organization must intimately understand both its customers and its abilities to manage its output.

With the caveat in mind that cost–benefits analysis does not provide the complete picture, it is worth restating that cost–benefits analysis is a useful, though often underused, tool for decision making in a research and development environment. It is not my intent to provide a full-blown treatise on either cost–benefits analysis, differentiation analysis, or computer-based tutoring in this chapter. As cost–benefits analyses are dependent on situational factors inherent to an application, subject-area domain, and organization, it is not possible to provide an all-encompassing cost–benefits analysis for a technology that is valid in all situations. Instead, this chapter offers a worked-through illustration of how a cost–benefits analysis might be performed to decide among strategies for implementing an intelligent tutoring system (ITS).[1]

In the following sections, examples of how to conduct a cost–benefits analysis when considering the adoption of computer-based tutoring technologies[2] are provided. Some of the major costs and benefits to be considered when evaluating tutoring technologies are examined. Finally, two hypothetical scenarios based on the capabilities of intelligent tutoring are provided to illustrate the approach.

COST–BENEFITS ANALYSIS

Cost–benefits analysis provides a financial basis for selecting among a set of actions or decisions. At the most simple level, cost–benefits analysis involves tabulating a set of costs and benefits associated with a set of competing options. When probability estimates regarding the likely outcomes of each option are considered (e.g., an outcome is estimated to have an X% probability of occurring), expected values for costs and benefits of each option can be computed. The option with the greatest value of the difference between expected benefits and expected costs is the option to choose.

---

[1]This chapter assumes that the reader is familiar with the various computer-based training technologies. For more information on this topic, see Anderson et al., 1986; Reiser et al., 1985; Wenger, 1987; also several chapters in this book.

[2]The term *computer-based tutoring* is used here in a generic sense to refer to any computerized tutoring system, whether knowledge-based or non-knowledge-based. Included are technologies such as intelligent tutoring systems, performance support systems, traditional computer-aided instruction, and the like.

Although cost–benefits models can be quite sophisticated, the author's preference is for a simple, easy-to-understand and easy-to-apply model. This bias towards simplicity has at least three advantages: (a) the cost–benefits analysis can be quickly and easily computed without expensive software (a spreadsheet works quite nicely); (b) the results of the analysis can be easy communicated to stakeholders (e.g., managers and clients) who may have little time or patience for an overly complex model; (c) the ease with which the analysis can be conducted is directly related to the probability that it will be used—the easier the analysis, the more likely it will be used. The author suggests using the model proposed by Hayes (1981), as follows:

A. For each of the alternative actions [or *options*, as described earlier]:
   1. Identify all the important sources of costs and benefits.
   2. Estimate the values of the costs and benefits.
   3. Estimate the probabilities of obtaining the costs and benefits.
   4. compare the expected values of the costs and benefits.
B. Choose the action for which the expected value of the benefits minus the expected value of the costs is greatest.

A key idea in cost–benefits analysis is that the *important* sources of cost and benefits should be identified. As Hayes (1981) observed, however, identifying costs and benefits can be problematic. Complex situations usually involve widespread effects that may not be easy (or possible) to identify before a decision is made. This is particularly true for situations in which organizations have little prior knowledge, such as intelligent tutoring.

Another difficulty is that of estimating probabilities. Psychological research is rife with examples of how decision making can go awry when people estimate probabilities (see, for example, Kahneman, Slovic, & Tversky, 1982). The accurate estimation of probabilities often requires more data than are available, especially in cases of new and untested technologies.

Problems also arise when comparing values of different types, such as the value of a job versus the value of a life. To illustrate this difficulty, Hayes (1981) used the following example: "Suppose you are offered two summer jobs as a bank guard at different branches of the same bank. One job in the suburban branch where there is zero risk to your life pays $5000 for the summer. The other job is in a busy urban area and involves a one percent risk to your life over the summer" (p. 191). How much more do you need to earn at the urban job to make up for the 1% increase in risk to your life? Is $100 enough, Hayes asked, or is $1 million necessary?

Although it may be difficult (or nearly impossible) for people to attach a value to probabilities related to loss of their life in general ("How much is a human life worth?"), it is often possible to gather enough information in specific cases (such as the previous example) to make cost–benefits analysis possible.

COST CATEGORIES

According to Fowler (1990), costs can be broken down into development costs and delivery costs. In each of these categories, it is possible to distinguish between overhead and capital costs. This leads to the breakdown displayed in Fig. 3.1.

Because the implementation of a new technology such as computer-based training, or even the revamping of a traditional lecture-only training course, will cost money "up-front" that can be amortized over the life of the training materials (or computer system), it is useful to identify costs associated with development and delivery separately. In addition, it is useful to distinguish between overhead costs and labor costs in each category.

Overhead costs have two components: labor and expense. Labor includes such costs as salaries, bonuses, vacation, and benefits paid to the employee. Expense costs cover such items as travel, office supplies, facilities, and rent. Capital costs tend to be big-ticket items (e.g., items over $1,000) such as printers, computers, and other equipment that have a useful life of over one year and are depreciated over the course of many years. The following sections examine some of these costs in terms of tutoring technology.

Training Development Costs

In this section, we look at major costs involved in designing and implementing a computer-based training system.

Workstation Costs (Capital). A workstation is the computer equipment used to develop a computer-based tutor. The workstation may include equipment other than the computer, such as a videodisk machine and monitor, CD-ROM drive, or a videotape machine, depending on the training system.

Because ITSs typically require high processing speeds, extensive memory sizes, and advanced user interface features (e.g., bitmapped displays, windows, mouse, etc.), they rely on workstations that are typically larger and more powerful than the personal computer setups commonly used in tra-

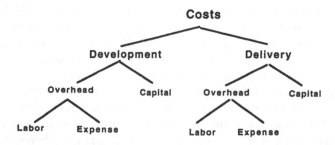

FIG. 3.1. Breakdown of financial cost categories (adapted from Fowler, 1990).

ditional computer-based instruction. As a consequence, ITS workstations are somewhat more expensive to run and maintain than personal computers (though this cost is rapidly decreasing to the personal computer level). In terms of a delivery workstation, costs to consider include:

- Hardware costs per student workstation (includes computer and any other auxiliary equipment such as a CD-ROM drive, memory, and display monitor).
- Software costs per workstation.
- Software and hardware maintenance costs per workstation.
- Workstation set-up costs (installation/hook-up of hardware and software).
- Costs associated with connecting the system to a network (if applicable).

System Development Costs (Overhead). System development costs vary widely from project to project, depending on the particular training needs. The primary cost to consider when developing automated training is the acquisition of personnel.

Unfortunately, it is rare to find individuals who have all the necessary skills; current technology does not provide authoring tools advanced enough to be used by an instructional designer who has no expertise in other areas of system development (although efforts are being made to develop such authoring tools for ITSs, such as the LEAP tutoring system at US WEST, described in this volume). For an ITS, each of the following skills, at a minimum, is required: knowledge acquisition, task analysis, knowledge-base development, instructional design, domain expertise, user interface design, system testing, and technical documentation. Although these skills are sometimes combined (e.g., someone good at knowledge acquisition may also be a good knowledge-base programmer, a good instructional designer may be a good technical documentation person, etc.), it is likely that an ITS development team will comprise at least five or six people.

As Sherman (1990) observed, custom development of training systems is usually competitive from a differentiation perspective, though not necessarily from a cost perspective. These observations lead to the conclusion that it may be cost beneficial to outsource (e.g., through a training vendor) for system development rather than develop the capability in-house until technical advances and availability of skilled personnel catch up with automated training needs. Whatever route is taken (contractors or full-time personnel), the associated costs should be included in the cost–benefits analysis.

An important source of cost relates to lesson development. This cost can be computed separately for the first lesson in a domain and for subsequent lessons. Costs for the first lesson in a domain are separated from subsequent development costs because system developers tend to become more efficient at lesson development as they learn to use authoring tools or training methods. As a result, subsequent lessons can usually be developed faster and with less cost than the initial lesson.

A common method of comparing lesson development is to look at *development–hour ratios*. A development–hour ratio expresses the amount of lesson development time that is required to produce one hour of training. Thus, a development–hour ratio of 100 indicates that it takes 100 hours of development to produce one hour of training. Typical development–hour ratios for the first lesson in a domain for different types of training are:[3]

- Instructor-led training: 60–70.
- Computer-aided instruction (text and simple graphics): 200–400.
- Self-paced training (e.g., with workbooks): 120–300.
- Intelligent tutoring system (with text and simple graphics): 400–600.

As these ratios indicate, development of traditional computer-aided instruction is closer to that required for self-paced training and IT development than for instructor-led training. The development–hour ratio of ITSs, which lies at the upper end of the traditional computer-aided instruction estimate, is attributable largely to the complexity of building a knowledge-based system. As authoring tools for developing ITSs (as currently exist for building nonknowledge-based training systems) become available, ITS development–hour ratios should fall to the level of traditional computer-aided instruction development.

Training Delivery Costs

We start by examining primary costs typically associated with delivery of training to the student population. Although most of these cost categories apply to training in general (including traditional noncomputer situations such as classroom training), in some cases, the cost categories refer specifically to computer-based training (these cases are noted in the text). The reader is reminded that these are not the only costs involved but, rather, are those that have been identified[4] as being important ones to consider.

---

[3]These estimates should be considered "rules of thumb" rather than hard and fast numbers—thus, development–hour ratios will differ according to the type of and complexity of the courseware being developed. The estimates provided here come from a variety of sources, including the author's personal experience and discussions with training organizations and individuals with whom the author has worked. See Barnes (1992) and Fulbright (1993) for additional information on development costs.

[4]Based on discussions with training personnel within US WEST and Bellcore. See also Anderson et al. (1985, 1986).

Training Overhead Costs. Off-site training, that is, training that takes place at a site other than where the employee normally works, is common. Regardless of the training media used, expenses commonly associated with off-site training include:

- Tuition cost per student.
- Room and board costs (when student must stay at a training site away from the normal workplace).
- Transportation costs to get to the training site.
- Long-distance communication costs (calls to family and office when student is in training away from home).

These costs can easily be estimated for most organizations and are usually available from a human resources or training department.

Productivity Costs Versus Labor Costs in Delivery. A common mistake in estimating costs is to make a claim for reduced labor costs when a new technology reduces training time. Because of reduced training time, the reasoning goes, less of the student's salary and benefits are paid out during training (during which time revenue is typically not generated by the student), resulting in a financial benefit. Even though there are often valid reasons for making a claim of reduced training time with computer-based technologies, claims for *financial benefits* related to labor costs during training should not typically be made.

Unless the student's salary and benefits are paid by the training organization or unless the organization pays the employee more than the normal salary while in training (both are unusual situations indeed!), the employee's home organization usually pays the employee at the same rate whether the student is in training or is a worker on the job. Reducing an employee's hours in training, then, does not usually save the organization any more in terms of labor than having the employee at work. As Fowler (1990) noted: "Never say the word 'save' in front of the Finance Department unless there are real dollars to be removed from the budget" (p. 17).

There is a clear difference, however, in terms of productivity between work and training situations: When an employee is in training, he or she is not usually contributing to the revenue of the home organization. Thus, costs and benefits related to employee *productivity* rather than employee salary should be examined. Lost productivity costs for training can be estimated by determining how much income the organization is losing by having the employee off the job. In the direct case, this can be estimated (after Fowler, 1990) by the following computation:

$$\text{widget value } * \text{ number of widgets not produced during training}$$
$$= \text{lost productivity cost.} \tag{1}$$

Another method for estimating lost productivity, which may be useful when it is not as clear how the employee contributes to organizational income, is to estimate the employee's daily value based on the organization's gross profit. This can be estimated (after Fowler, 1990) as:

gross profits / annual work days / number of employees *
class days * number of students = lost productivity. (2)

This second approach is a "soft" method and assumes the organization makes a profit. In some organizations, such as research and development, this may not be true. In that case, an estimate based on how much it costs to replace an employee in training can be estimated (after Fowler, 1990) as:

temporary daily cost * number of employees replaced *
number of class days = lost productivity. (3)

Although these three approaches vary in terms of how they estimate productivity loss, the selection of a method should be driven by the available data (some local organizations may be very reluctant, or not willing at all, to release productivity data) as well as the type of organization (e.g., research and development, profit center, etc.).

Capital Costs in Training Delivery. Delivery of computer-based training need not be any more expensive than traditional lecture-only training. Costs incurred by using computer technology to deliver on-site, self-paced training (e.g., data transmission costs, hardware maintenance costs) are often offset by reductions in instructor costs, lost productivity, and travel expenses incurred by traditional, off-site training.

The primary capital costs in delivery of computer-based training are similar to those for development: hardware costs associated with maintaining and replacing computers and other training equipment (e.g., CD-ROM drives, etc.). Depending on the training system delivered (e.g., whether it requires video, memory and disk storage requirements, whether an advanced workstation is required or just a personal computer, etc.), hardware costs will vary widely but must be estimated for the cost–benefits analysis.

BENEFITS CATEGORIES

Financial benefits refer to actual savings in terms of increased revenues or decreased costs. These benefits are distinguished from the differentiation advantages just discussed in that financial benefits are expressed directly in terms of dollars (if differentiation benefits can be quantified, they can be expressed as financial benefits). Although issues such as student motivation, performance levels, and other difficult-to-quantify benefits are extremely important to consider, focusing on financial benefits is absolutely necessary if the goal is to convince management that a new technology such as ITS is a worthwhile investment.

Kearsley (1982) identified three categories of financial benefits. Any justification should demonstrate one of these three types of benefits: efficiency, effectiveness, and productivity.

Efficiency benefits are demonstrated by a reduction in costs while output remains the same. In the case of computer-based tutoring, efficiency is demonstrated by reducing the overall training costs while maintaining student performance levels on the job. For example, a tutoring system might provide on-site training (thus reducing travel expenses and instructor time) and produce equivalent on-the-job performance as an original off-site lecture course. In this case, the tutoring system has reduced training expenses while maintaining the previous level of training efficacy, which is cost efficient.

Cost *effectiveness* is demonstrated when costs are held level and output is increased. For example, a computer-based training system that is comparable in terms of development and delivery costs (amortized over the life of the training program) to a traditional lecture-only course but raises student performance levels, compared to the lecture-only course, is cost-effective.

Productivity is demonstrated when costs are decreased while output is increased. A computer-based training system that costs less to deliver over the long run than a traditional lecture course but also increases student performance is cost-productive. As a result, demonstrating cost productivity is the ultimate financial justification for adopting a training program.

Different computer-based training programs fall into different benefits categories. The fit depends on the specifics of the training situation (e.g., the domain, the student audience, the organizational goals and capabilities). In general, however, claims for traditional, computer-assisted instruction tend to be in the cost-efficiency department: Learning levels similar to traditional training techniques (e.g., lecture, self-paced) can be attained at reduced cost. Typical claims made for ITSs, on the other hand, tend to fall into the cost-effectiveness and cost-productivity categories. This reflects the goal of ITS developers to increase student competency by providing training and feedback directly relevant to the skills needed in the target domain.

Expected Performance Improvements

Tutoring systems are typically integrated with existing training (e.g., classroom lecture, self-paced study) as a way to augment student skills and knowledge.[5] As a result, training benefits for ITSs are usually expressed in terms of student-performance gains when tutoring systems are used in conjunction with traditional classroom training, compared to classroom training without augmentation of tutoring systems.

Some performance improvements noted in the literature for intelligent tutoring systems include (see Anderson, Boyle, & Reiser, 1985; Anderson & Lesgold, 1986):

• Reduction of performance errors by 50%. In a business setting, this translates into reduced costs to correct errors (e.g., reduction in the number of errors in new orders in service provisioning).

• Higher student skill level after training, specifically, a two standard deviation[6] improvement in performance. An industry standard for converting improvements in performance into dollar amounts is to equate every one standard deviation performance into productivity gains equivalent to 40% of an employee's salary (Godkewitsch, 1987). Thus, a two-standard deviation improvement could be estimated as productivity gains equivalent to 80% of the employee's salary.

• Reduction in time to reach performance criterion by 50% (e.g., reducing training time from 8 weeks to 4 weeks). This reduction translates into lower productivity costs as the student is away from the job for a shorter time.

Expected Savings

Several potential cost benefits can be realized from using computer-based training, such as:

• On-site training reduces costs associated with travel to distant training sites.

• Automated training can reduce training time overall (reduced time-to-criterion).

• Instructor time and associated costs are reduced.

• Time spent using specialized training facilities may be reduced (depending on where the training is delivered).

---

[5]There may be a variety of reasons for using computer-based tutors to augment rather than replace traditional training, ranging from organizational unfamiliarity with training technology that could potentially replace entire classroom sessions (e.g., intelligent tutoring), to the intentional use of systems that can, at best, augment but not replace other forms of training (e.g., traditional computer-aided instruction).

[6]A standard deviation is a measure of dispersion about an average score. In a normal distribution of student scores (normality can be assumed when there are at least 20 scores), about 68% of the scores fall between −1 and +1 standard deviations from the average score. Only 2% of students fall 2 or more standard deviations from the average score.

- Distribution of new courseware can be facilitated in a computer environment.

- Estimation of some of these costs, such as reduced travel and instructor costs, is straightforward. Other savings are more difficult to estimate before a system is actually built and tested, such as those related to reduced time-to-criterion and courseware distribution costs. In these cases, rules of thumb (e.g., as those noted in the previous section) must be used to estimate savings.

Differentiation Advantages

To review, differentiation advantages describe nonfinancial benefits associated with a technology (e.g., student motivation, ease of use, fit to the organizational culture, etc.).

Differentiation benefits are difficult, if possible, to quantify economically because data may be difficult to obtain or because quantitative indicators for the benefits may not exist. The important point, though, is that employees, managers, customers, trainers, and all other personnel involved in training may often perceive value in using computerized training, and this value may translate into a variety of differentiation advantages, such as:

- Students like ITSs. This translates into student motivation to learn.

- Because computerized training is self-paced, students have a feeling of control. This translates into an *internal locus of control* over the learning environment in which students are learning because they want to learn rather than because an external source, such as an instructor, is compelling them to learn.

- The self-paced nature of computer learning allows the students to make mistakes and review material without being embarrassed by their performance. As a result, students may spend more time acquiring critical skills than they would in a more "public" learning environment such as a classroom.

- Computer delivery of training allows the students to practice and learn new information when appropriate for them (not just when an instructor and other students are ready).

- Managers may be happy to have workers on or near the job site during training.

- Students may be happy to train near home and not have to travel to distant sites for extended periods of time.

- Computerized training allows a consistency of training and performance assessment across all students.

- Automated training may "fit in" with an overall technology plan to develop intelligent user environments in a workplace (e.g., using ITS to train users at employees' workstations).

- ITSs permit flexibility in training: Instructors can combine lecture and ITS practice simultaneously with different student groups; students can practice on their own time, and so on.

- In customer-contact jobs, increased productivity and expertise of the employees should result in higher customer satisfaction (customers are assisted more quickly and with greater skill than with previous training).

In some cases, it is possible to quantify the expected benefits from improving areas such as customer service (e.g., there may be a predicted quantitative relationship between customer service metrics and revenue). In that case, this information should be used in the cost–benefits analysis.

ANALYSIS SCENARIOS

In this section, two training scenarios are examined. In each case, training costs and benefits for a hypothetical job are discussed. These scenarios are provided as examples of how to estimate costs for training situations.

For this illustration, the job of a customer service representative has been selected for analysis. Our hypothetical customer service representatives take orders for products and services from customers who call them over the phone. This is an essential business function, found in any business that provides a phone number for customers to call for service, such as credit card companies, mail-order businesses with phone order lines, and telephone companies, to name a few.

In these training scenarios, some standard costs that might go into training customer service representatives are examined. We assume that this hypothetical training situation has the following characteristics:

- There is high employee turnover, requiring hundreds of students to be trained each year.
- Training is extensive—25 work days of classroom training.
- Training is delivered at central sites to which many students must travel because they do not live within commuting distance.

Given the particulars of this training situation, our imaginary company has decided to investigate the cost of implementing an ITS as a means of providing local training (i.e., at the students' workplace) for at least part of its 25-day course, if not for all of it. In addition, management is hoping for cost-effective benefits with an ITS (though cost-productivity is welcome!)—that is, it is hoped that an ITS training system will result in improved on-the-job performance at costs similar to that of traditional training.

The first step in performing a cost–benefits analysis for this situation is to look at current training costs. In the next sections, two scenarios are examined: In Scenario 1, the costs and benefits of a traditional method of training (classroom) are estimated. In Scenario 2, customer service representatives receive training from both lectures and an ITS that is located at the workplace.

Scenario 1. Do Not Change Current Training Methods

New hires travel to central training locations at major cities for a 5-week (25-day) training course. This course combines classroom lecture and traditional computer-aided instruction (with classroom experience predominating). The company trains about 525 new hires a year (this is a high-turnover job).

Assumptions. The following set of assumptions is used in this scenario:

- Students travel to a regional training site for 5 business weeks of classroom and traditional computer-based training.
- On average, trained service representatives produce approximately $120,000 in sales revenue per year, which translates to about $2,400 in sales for every week on the job.
- Costs are incurred for students' room and board, time, with one visit back home (if from out of town).
- Company trains 525 new employees per year.

Benefits. In our example, trained employees account for approximately $120,000 in sales per year. This is the expected value (EV) of the benefit per student per year: EV (benefits) = $120,000.

Steady-State Costs. Costs that are predictable and steady in the sense that the organization can count on these costs every time a student is sent off-site for training include:

- Food: 25 days @ $40/day = $1,000.
- Lodging: 25 days @ $75/day = $1,875.
- Travel: $1,000.
- Course tuition: $5,000.
- Lost productivity for 5 weeks @ $2,400/week = $12,000.

The total, steady-state cost is the sum of these values: $20,875 per student.

Because these costs are known to occur in training, their probability of occurrence is 1.0. Thus, the expected value (EV) of costs per student per year is: EV (costs) = $20,875.

Start-up Costs. These are costs associated with starting up a training course. Because employees are sent to a training center at which a course has been established, there are no direct start-up costs for the employees' organization (these costs are borne by the training organization and are passed along in terms of tuition). These costs total $0.

Analysis. Following the method just outlined, we should compute a cost–benefits value by subtracting the expected value of the costs from the expected value of the benefits. In other words:

EV (benefits) − EV (costs) = ($120,000 − $20,875) = $99,125.

The financial benefit of sending an employee to training (Scenario 1) is estimated to be $99,125 per employee. This result will be compared to the implementation of an IT in Scenario 2.

Scenario 2. Develop an ITS That Will Improve Time-to-Criterion by 30%

In this scenario, it is assumed that a standard measure in performance training, *time-to-criterion*, can be reduced by 30% for each student by introducing the use of an ITS in training. Time-to-criterion refers to the time it takes a student to reach a minimum level of competence. Thus, reducing this time by 30% allows an overall reduction in training time of 30%—that is, from 25 days to 17.5 days. This reduction in training time translates into decreased productivity losses as the employee is away from the workplace for a shorter time. This level of improvement is within the reported parameters for ITSs in academic settings (see Anderson et al., 1985; Anderson & Lesgold, 1986).

Another reason to train with ITSs is to improve student performance. Although few controlled studies have been done in which students are compared based on training in traditional settings versus training with ITSs, at least one investigator (Anderson & Lesgold, 1986) reported improvements of two standard deviations when an ITS was included as part of classroom training compared to a situation in which classroom training did not include an ITS.

A natural assumption in a business setting is that performance improvements will translate into financial benefits, perhaps because the well-trained employee is more efficient at placing new orders, makes fewer mistakes, and keeps customers happier than less-well-trained employees. The problem is translating these benefits into dollars. One rule-of-thumb was provided by Godkewitsch (1987) who reported a study in which industrial psychologists equated one standard deviation improvement in performance with 40% of an employee's annual wage. In other words, an employee who performs at less than one standard deviation below the average is worth 40% less than an employee who performs at the average level. Likewise, an employee who performs one standard deviation above average is worth 40% more than an average performer.

For the present scenario, we target an on-the-job performance improvement of one standard deviation over what can be obtained with the traditional training provided in Scenario 1. Again, this is within the realm of reported results for ITSs.

Finally, the ITS is delivered either (a) by embedding it in the same workstation environment that the service representatives use on the job or (b) by providing a stand-alone workstation (separate from the workstation used on the job) at the employees' place of work. This eliminates the travel expenses to an off-site training location. The ITS is used *in conjunction* with standard classroom training given on-site. The ITS could be used to present new material and to allow the student to practice certain concepts and skills in a self-paced mode.

Assumptions. The following assumptions are made for Scenario 2:

- A 30% reduction in time-to-criterion (within reach of ITSs) compared to Scenario 1 is targeted.
- Reducing training by 7.5 days reduces productivity losses accordingly (i.e., the saved days would be spent on the job).
- Service representatives earn an average annual salary of $35,000.

Outcome Cases. A technical review of the situation identified three possible outcomes. As a result, probabilities were attached to each outcome:

- *Case 1:* The ITS will not work and will not produce any improvement in time-to-criterion measures or performance measures. This probability is judged to be 0.10.
- *Case 2:* The ITS will perform as planned and will produce the expected time-to-criterion improvements of 30% and improvement in on-the-job performance of one standard deviation. This probability is judged to be 0.70.
- *Case 3:* The ITS will work only half as well as expected and will produce time-to-criterion improvements of only 15% and a 0.50 standard

deviation in on-the-job-performance. This probability is estimated at 0.20.

Benefits. Benefits can be computed as savings in terms of reduced productivity loss as the training time is reduced. Given an estimated productivity level of $2,400/week (as estimated in Scenario 1), this translates into productivity of $480/day for service representatives on the job. This figure will be used in computing benefits under each of the outcome cases.

Because training is now done in-house, expenses related to food, lodging, and travel are eliminated, thereby saving $1,000 + $1,875 + $1,000 = $3,875 per student (using the figures from Scenario 1).

We use the assumption (after Godkewitsch, 1987) that one standard deviation improvement in performance is worth 40% of the student's salary. Thus, a one standard deviation improvement for a service representative who makes $35,000 per year is worth: 0.40 * $35,000 = $14,000. Similarly, an improvement of 0.50 standard deviations would be worth (0.40 * 0.50 * $35,000) = $7,000.

To compute expected values for Scenario 2, it is necessary to consider the probabilities associated with each outcome case cited above. Specifically:

1. In outcome case 1, there are no savings attributable to the ITS because it does not work (i.e., it does not reduce time-to-criterion). The expected value (EV) of the benefits of outcome 1 are the probability of its occurrence multiplied by its benefits, or:

EV (benefits of outcome 1) = 0.10 ($0) = $0.

2. In outcome case 2, the ITS reduces the course length by 30% with probability 0.70. The expected value (EV) of the benefits in this case is a 7.5 day increase in productivity (reducing training by 30% removes 7.5 days), a $14,000 gain in on-the-job performance over the $120,000 from traditional training, and the savings realized by conducting the training in-house. Thus,

EV (benefits of outcome 2) = 0.70 [(7.5 * $480) + $134,000 + $3,875] = $99,032.50.

3. In outcome case 3, the ITS reduces the course length by 15% with a probability of 0.20. As a result, training is reduced by only 3.75 days. Benefits are computed as productivity gains for 3.75 days, on-the-job performance gains of $7,000 over the $120,000 from traditional training, and savings from training in-house:

EV (benefits of outcome 3) = 0.20 [(3.75 * $480) + $127,000 + $3,875] = $26,535.

Summing the EVs from all three outcome cases, we have:

EV (benefits of Scenario 2) = $0 + $99,032.50 + $26,535 = $125,567.50.

Steady-State Costs. In Scenario 1, ongoing costs for sending students to training were based primarily on travel expenses and tuition. When training is done in-house with computers, steady-state costs revolve more around maintenance of the training equipment and facilities. The following are costs the company has identified for delivering an ITS course (costs are expressed per student):

- Facilities rental: $30/student/training day.
- Computer workstation maintenance: $200/student.
- Lost productivity: $480/day.

Start-up Costs. These are costs related to starting a new training program. Estimates for developing ITS training vary widely and are highly dependent on the specifics of the training situation. Let us assume, however, that the company has done a study and estimates that it will cost $2 million to develop and deliver an ITS on 20 workstations.[7] In addition, this ITS is expected to serve a useful training life of 20 years (the job is not expected to change significantly over that time). Spread over 20 years, the start-up cost is $100,000 per year for the life of the training system. Assuming 525 students per year, this is a cost of approximately $200.00 per student.

In addition, it is necessary to hire an instructor to assist students in using the ITS and to provide basic lecture material (remember, the ITS augments the instructor; it does not replace him or her). Assume the instructor charges $500/day for training services and that there are 25 students per class, with 21 classes per year. For a 25-day course, this is a per-student cost of $500 for the instructor. For a 21.25-day course, the per-student instructor cost is $425, whereas for a 17.5-day course, the per-student instructor cost is $350.

This leads to the following costs analysis:

1. For outcome case 1, the ITS does not work with probability 0.10. Thus, costs are computed based on facilities rental, workstation mainte-

---

[7]This cost, like all the others in this report, is entirely imaginary—fabricated only for this example!

nance, and lost productivity as well as the start-up costs for a 25-day course:

EV (costs of outcome 1) = 0.10 [(25 * 480) + (25 * 30) + $200 + $500 + $200] = $1,365.

2. For outcome case 2, the course is reduced in length by 30%, resulting in a 17.5-day course, with probability 0.70. Thus,

EV (costs of outcome 2) = 0.70 [(17.5 * $480) + (17.5 * 30) + $200 + $350 + $200] = $6,772.50.

3. For outcome case 3, the course is reduced in length by 15%, resulting in a 21.25-day course, with probability 0.20. Thus,

EV (costs of outcome 3) = 0.20 [(21.25 * 480) + (21.25 * 30) + $200 + $425 + $200)] = $2,332.50.

Summing the expected costs from the three outcomes with the system development and deployment costs to arrive at a single expected cost for Scenario 2, we have,
EV (costs for Scenario 2) = $1,365 + $6,772.50 + $2,332.50 = $10,470.

Analysis.

EV(benefits for Scenario 2) − EV(costs for Scenario 2) = $125,567.50 − $10,470 = 115,097.50.

Thus, for the second scenario, in which an ITS is developed to augment classroom training, we have estimated the net benefit per student to be $115,097.50.

COST–BENEFITS COMPARISONS

In the aforementioned cost analyses, the following results were obtained:

Scenario 1: Do not change the way training is currently done:
EV(benefits) − EV(costs) = $99,125;

Scenario 2: Build an ITS to reduce time-to-criterion by 30%:
EV(benefits) − EV(costs) = $115,092.50.

Following the rule that we should choose the option for which the EV(benefits) − EV(cost) value is the largest (see Fig. 3.1), we should adopt a project to meet the goals of Scenario 2—build an ITS that reduces time-to-criterion by 30% and improves on-the-job performance by one standard deviation over traditional training.

SUMMARY AND CONCLUSION

In this chapter, a sample cost–benefits analysis for an ITS was illustrated. Several simplifying assumptions were made, specifically:

- That the costs of initial system development and deployment could be accurately estimated up-front.
- That there were only two criteria for justifying the effectiveness of a tutoring system (time-to-criterion and on-the-job performance, when in fact, there are several that could be considered).
- That success/failure probabilities could be accurately estimated.
- That personnel capable of building such a system are available and are essentially "cost-free" (i.e., they are employees of the company and do not charge extra to build an ITS over a traditional course).

Any or all of these assumptions may not be true for the reader's organization or situation and should be modified accordingly when performing a cost–benefits analysis.

In addition, these examples did not quantify benefits that might be useful to consider in a real situation. These include such benefits as student motivation, using ITSs, and improved on-the-job performance, using ITSs. A difficulty, however, is that it may not be possible to quantify these factors until (or unless) data have been collected to measure the benefits with similar systems.[8]

In an actual example, there are a number of additional costs and benefits to consider: Specifics of the domain, the organization, the students, competitive differentiation and cost considerations, and a variety of other situational factors contribute to the final cost–benefits analysis. In attempting to estimate costs, benefits, and probabilities, much can be learned about the current situation and what is needed to improve it, and this is an exercise guaranteed to improve the quality of decision making.

REFERENCES

Anderson, J. R., Boyle, C. E., & Reiser, B. J. (1985). Intelligent tutoring systems. *Science, 228*, 107–131.

---

[8]To the author's knowledge, no attempt has been made to quantify these factors in other published ITS research reports.

Anderson, J. R., & Lesgold, A. (1986). Tutorial on intelligent tutoring systems delivered at the *American Association for Artificial Intelligence*.

Barnes, D. (1992, June). The economics of computer-based training. *Conference record for the 1992 IEEE fifth conference on Human Factors and Power Plants* (pp. 422–431). Monterey, CA.

Fowler, W. R. (1990, Fall). Cost justifying your training project. *Journal of Interactive Instruction Development*, 16–19.

Fulbright, T. W. (1993, May). The computer-based job aid: An effective alternative to traditional factory training. *Proceedings of IEEE 1993 National Aerospace and Electronics Conference* (pp. 1021–1027). *NAECOM.*, Dayton, OH.

Godkewitsch, M. (1987, May). The dollars and sense of corporate training. *Training*, 79–81.

Hayes, J. R. (1981). Cost–benefits analysis. In J. R. Hayes (Ed.), *The complete problem solver* (pp. 183–195). Philadelphia: Franklin Institute Press.

Kahneman, D., Slovic, P., & Tversky, A. (1982). *Judgment under uncertainty: Heuristics and biases*. New York: Press Syndicate of the University of Cambridge.

Kearsley, G. (1982). *Costs, benefits, and productivity in training systems*. Reading, MA: Addison-Wesley.

Reiser, B. J., Anderson, J. R., & Farrell, R. G. (1985). Dynamic student modeling in an intelligent tutor for LISP programming. *Proceedings of the ninth Joint International Conference on Artificial Intelligence*, Los Angeles, CA, 8–14.

Sherman, C. (1990, August/September). Buying smart: Training technology and competitive advantage. *Bulletin of the American Society for Information Science*, pp. 14–15.

Wenger, E. (1987). *Artificial intelligence and tutoring systems*. Los Altos, CA: Kaufmann.

II

CASE STUDIES FROM INDUSTRY

4

INTRODUCING ADVANCED TECHNOLOGY APPLICATIONS INTO CORPORATE ENVIRONMENTS

Charles P. Bloom
A. Scott Wolff
Brigham Bell
Applied Research and Multimedia Services
US WEST Advanced Technologies

Training has traditionally been viewed in industry as primarily a cost-centered activity. In the short term, training consumes revenue; it does not generate revenue. As a result, training is typically seen as a last-ditch response to a crisis such as a shortage of employees with critical skills. Although it is certainly true that corporate managers recognize that training is often necessary to bring employees to a level that allows them to produce revenue, the traditional bias toward keeping training at the absolute minimum remains prevalent. The numbers are even more depressing: In 1992, it was estimated by the American Society for Training and Development that American companies spend in the neighborhood of $30 billion a year on training. Even though this number at first seems impressive, further examination shows that this expenditure represents spending on a tenth of the workforce by less than 1% of American companies. Additionally, most training dollars are spent on managerial positions and not on "frontline workers" ("Training and the Workplace," 1992).

It is also clear that the model of how to educate and train has not been upgraded to reflect the technology of the times. Although computer laboratories are becoming increasingly popular as classroom tools, the comparatively low capabilities of many computer-based training systems (which are often little more than expensive, computerized page-turning devices) have relegated this technology to a "second class" status. These tools are used primarily as a way to offer extra material to interested students or to augment lecture-based instruction. Even though there has been movement towards "on-demand" training such as instructional television (Yeager, 1991), the model of the student as primarily a passive participant in the training environment persists. As a result, the introduction of advanced, interactive computer-based learning environments must successfully counter these traditional views of training to be adopted.

These cost-centered, low-technology, crisis-based views of training are gradually being replaced in public schools, universities, corporate, and government training rooms (e.g., "Coping," 1994; DeMarco, 1990). As one writer observed, "It is no easier to be against training and education for America's adult workers than it is to be against smallpox jabs for children" ("Training and the Workplace," 1992). With these shifting views, a renewed interest in the use of technology for training delivery is on the rise. One forecast predicted that by 1997, 18% of corporate training budgets will be specifically targeted at computer-based training ("Computer-Based Learning," 1994). Although this trend will undoubtedly continue to be pushed by decreasing computing costs and increasing computing power, there have been recent political and societal changes that have reenergized interest in technological approaches to training delivery.[1] In fact, the general public seems more willing to devote personal income to network services that involve education than to any other type of service, placing it above entertainment as a potential marketplace commodity (Piller, 1994). In industry, re-engineering has been a dominant force in driving training as an essential component of supporting large-scale system and business process improvements.

At present, technological, political, and economic forces are shaping how training should and can be delivered in industry, public schools, universities, and government, and are producing interactions among these sectors. The military and various government agencies have long been proponents of, and major funding sources for, research and development in advanced, electronic, interactive learning environments. More recently, industrial companies such as US WEST have begun to pursue technology to benefit training delivery in the public school systems (see also DeMarco, 1990) as well as their own internal training programs (Bloom, Bell, & Linton, 1994; Bloom, Bell, Meiskey, Sparks, Dooley, & Linton, 1995).

The focus of this chapter is primarily on the corporate environment, as that is where our experience lies (though we anticipate that our observations and suggestions will have relevance to other types of organizations, whether civilian or military, commercial or public). Specifically, we discuss

---

[1]For example, one has only to visit a bookstore to see the proliferation of books on Internet use that has appeared during the last year or so. In addition, proposals for educational technology, starting with NREN under the Bush administration, and continuing with specific initiations such as Project Globe under the Clinton administration, are illustrative of these influences.

barriers to introducing advanced, interactive learning environments (such as ITSs) to a corporate organization, and we discuss factors critical to success in such an endeavor.

In the following sections, we address how to sell the concept of advanced technology training delivery to internal management, how to design and develop advanced technology training applications, and how to successfully deploy these applications in the corporate training environment.

SELLING ADVANCED TECHNOLOGY TRAINING APPLICATIONS TO MANAGEMENT

The best technology in the world, however well suited to your organization's needs, will not be implemented unless all affected managers (at all levels) recognize its value. This chapter adopts as its central focus an ITS developed at US WEST. This tutoring system, dubbed LEAP (Learn, Explore, and Practice), was designed to provide customer-contact employees (CCEs) with an intelligent, coached apprenticeship environment to practice how to interact with customers and internal software systems to provision telecommunications products and services. LEAP is one of the few ITSs to be used on a wide scale in industry.

The initial phase of the LEAP project spanned approximately 3 years from conception to first prototype. During the period prior to prototyping, the dual concepts of initiating research and development on interactive learning environments (ILEs) and the notion of introducing an ITS for a specific training need were socialized within the US WEST technical and management communities. We are not trying to make the claim that this process was smooth and seamless; rather, it took far too long. However, with the benefit of hindsight, we can see what worked and what did not. In this section, we use our experiences with LEAP as a context for discussing strategies that can be used to introduce advanced training technologies into a corporate organization.

Identifying a Domain With a Business Impact

In a client-driven technology organization, the first order of business is to identify a clear client need for your system. This is not to advocate searching for a problem in need of an already-developed solution but rather to emphasize the need to identify a match between a valid training need and the ability to provide an effective technical solution. Certainly, not all training situations require electronic delivery of training, but those that can benefit from such can show impressive improvements in training quality for an organization. At US WEST, we quickly identified customer contact as a primary business activity and, therefore, one worthy of examination for potential improvement. After some study of job functions and training, it became apparent that the role of the customer contact employee (CCE) requires more training than initially expected (10–12 weeks of classroom training) and has a longer on-the-job time to proficiency than initially expected (6–18 months).[2]

Based on our observations, we were able to identify several reasons for these long training sessions and time-to-proficiency intervals. First, the job required sophisticated multitasking skills. CCEs had to simultaneously converse with customers to solve problems and sell services, manipulate phone calls, and correctly use a number of disparate, poorly integrated and outdated software systems. Second, each of these systems had numerous screens and fields that had to be understood and navigated during customer contact. In addition, the user interfaces of these systems could best be described as nightmarish—often, there was no clear way of distinguishing where fields were located on a screen and what the form of their content should be. Finally, CCEs needed to understand all features and incompatibilities of offered services, including information concerning their availability, price, and capabilities (which can differ from state to state, a considerable problem in a company that spans 14 states as does US WEST).

Armed with a strong knowledge of our clients' business needs and job functions, we were able to address the issue of training from an informed position. As we had no intention (much less capability) of replacing the workforce responsible for training the thousands of CCEs at US WEST, we identified the training organization as our next-process client (with the CCE organization as the ultimate client). At this point, we had three stakeholder groups to address: our own internal management, the next-process client organization (the learning group), and the ultimate client group (the CCE organization). In the next subsection, we describe how we addressed their needs.

Establishing Business and Technology Cases

We addressed the needs of our client stakeholder groups by conducting two feasibility studies: One focused on a technology analysis, whereas the other examined cost–benefits tradeoffs of employing an ITS for training in the CCE-task domain. Although this approach proved to be time consuming, it also proved to be effective.

Even though all three client groups (our technology organization, the corporate training organization, and the user population's organization)

---

[2]This is the amount of time it takes for customer service representatives to assess themselves as being proficient (where proficiency means that the employee can confidently handle most calls without relying on support from a more experienced person).

were receptive to both the technology and business cases we provided, we saw somewhat uneven interest among these groups. Ironically, the cost–benefits analysis created the most interest in the technology organization, but the two business organizations seemed more interested in the technology analysis. Each of these analyses is described as follows:

Technology Analysis. The primary output of our technology analysis was a feasibility report for our clients. The purpose of this report was to present the results of an analysis of various options for the client groups (of which IT was simply one of three viable options). Specifically, this report addressed the following issues:

- A comparison of traditional computer-based tutoring (CBT) with IT.
- A survey of training vendors.
- An analysis of the target domain (i.e., service provisioning).
- Options for computer-based training for service provisioning.
- Projected project timelines and staffing requirements.

Traditional computer-based training was compared with IT in the context of a structure proposed by Aronson and Briggs (1983). These investigators list nine instructional events that an effective tutor should address:

- *Gaining attention.* It is necessary to gain the student's attention to instruct.
- *Inform the learner of the objective.* It is necessary to tell the student what he or she will learn in such a manner that the student clearly understands the objective of the lesson.
- *Stimulate recall of prerequisite material.* The learner must be reminded of previous learning so that there is a basis for new learning.
- *Present the stimulus material.* A variety of media may be used to present the material to be learned.
- *Provide learning guidance.* The student should be guided toward the learning objective. This means that the learner must understand the concepts and/or skills involved in the lesson, and the tutor must be able to provide direction to the student when needed.
- *Elicit the performance.* The learner should perform an overt activity to demonstrate that learning has occurred.
- *Provide feedback about performance correctness.* Feedback must be specific enough to promote changes in the student's behavior.
- *Assess the performance.* The tutor must determine if the student has acquired the target concepts/skills.

- *Enhance retention and transfer.* The tutor should help the student retain skills/knowledge and transfer them to other situations.

This model was selected because it is "generic" in the sense that it can be applied to any tutoring situation, whether delivered traditionally or on a computer. On the basis of these nine criteria, traditional computer-based training was viewed as being less effective than IT in the areas of providing learning guidance, eliciting performance (or more specifically, in the area of evaluating the student's demonstration of performance), providing feedback about performance, and assessing the performance.

A domain assessment was also performed as part of the feasibility analysis. In this assessment, a number of attributes that should be present for an automated tutoring application were examined (by *automated tutor* we mean any type of training system).

- *High student flow exists.* To justify the expense of an automated tutor, there should be a high student turnover rate.
- *Critical skills and knowledge must be taught.* The domain must be sufficiently rich to justify the expense of an automated tutor (this is especially true for an ITS in which the domain must be sufficiently rich to warrant development of a knowledge base).
- *Training at remote sites is needed.* The need to train at student locations lends support in favor of an automated tutor.
- *Instructors are available, though not overly so.* When few instructors are available, automated tutors can both preserve the knowledge of those individuals and increase the coverage that the instructor can provide.
- *Project has high visibility.* The selected training application must have high visibility, particularly when automated tutoring is first introduced into an organization. This means that the system is addressing a training need for an important business function, and/or an area in which skilled employees are in short supply. This will increase the probability that the tutoring system gets the support and attention it needs to be successful.
- *Recurrent training is needed.* The greater the extent to which refresher or on-the-job training is required, the greater the justification for an automated tutoring system.
- *Training in the area is well understood.* It is difficult to build a tutor for domains in which knowledge is in short supply.
- *Subject-matter experts are available.* It is not enough that domain experts and instructors exist; these people must be made available to work on the development of the tutoring system.

- *Experts are able to articulate knowledge.* Domain experts must be able and willing to articulate their domain and pedagogical knowledge for inclusion in the tutoring system.
- *Expert knowledge can be represented.* To build an automated tutor, especially an ITS, the relevant domain and pedagogical knowledge must be representable in the tutor.
- *The student audience is well defined.* The target audience of tutor users and their needs must be understood and designed into the tutor.

Results of the domain assessment indicated that all of these attributes were present in the service provisioning domain and that all but one (instructor availability) were present to a high degree—instructor availability was assessed to be moderate. Therefore, we determined that the domain of service provisioning was suitable for automated tutor development. Further analysis of the domain into specific job tasks (e.g., sales negotiation, provisioning system use, complaint handling, etc.) led to the assessment that an ITS (rather than a traditional CBT) was required to teach many of these complex human–human and human–computer interaction skills.

Finally, our feasibility analysis provided two viable options for automated tutoring in the domain of service provisioning: (a) multimedia computer-based training for specific job tasks. This system makes use of a multimedia user interface (e.g., graphics, animation, video, as appropriate) to illustrate and drill customer contact employees in basic knowledge required for service provisioning. The tutor addresses such topics as what telephone line features/services are available for sale to customers, how these features work, how the features could be bundled into packages (following merchandising strategies), and some common sales techniques and tips. This option provided the advantages of using off-the-shelf technology and available training vendors for system development; (b) development of an ITS to teach consultative sales and provisioning system use. This ITS goes beyond simply teaching product/service facts and a few sales negotiation "tips" (though such training would likely be integrated in the ITS); it provides intelligent coached practice with human–human and human–computer interaction skills. Of the two options, this one was clearly the highest risk, but it also offered the greatest potential for providing effective training.

Based on these analyses, it was apparent that developing an ITS for service provisioning was the best choice. The next step was to conduct a cost–benefits analysis to determine whether the anticipated costs and risks of developing an ITS could be justified.

Cost–Benefits Analysis. Although we have separated "technology" and "business" issues in this chapter, the actual division between these areas is not that clear in practice. To conduct a cost–benefits analysis, it is necessary to have a good understanding of both IT technology as well as how to perform an effective cost–benefits analysis.

A key idea in cost–benefits analysis is that the primary sources of costs and benefits should be identified. Unfortunately, identifying costs and benefits can be problematic, as we discovered. We were unable to find prior work that clearly identified sources of costs and benefits (though these were certainly distributed in the ITS literature). At that time (1991), the best leads we could find were in the general area of training (e.g., Fowler, 1990; Godkewitsch, 1987; Kearsley, 1982). Given this paucity of prior work in the area, we developed our own cost–benefits model based on industry standards for computer-based training development (obtained from our corporate training organization) as well as anecdotal information and our own estimates regarding ITS development parameters.

Using an approach to cost–benefits analysis described elsewhere in this book (see chapter 3), we posed two scenarios for training: (a) Do not change current training; (b) develop an ITS that improves student time-to-criterion by 30%. For the ITS scenario, we postulated three outcome cases:

1. The ITS does not work and does not produce an improvement in time-to-criterion performance measures. This probability was estimated to be 10%.
2. The ITS performs as planned and produces the expected time-to-criterion improvement of 30% and on-the-job performance improvements of one standard deviation. This probability was estimated to be 70%.
3. The ITS works only half as well as expected and produces time-to-criterion improvements of 15% and on-the-job performance improvements of 0.5 standard deviation. This probability was estimated to be 20%.

For each of these three outcome cases, we computed expected values of costs and benefits. Summing the expected costs and expected benefits for the three ITS scenario outcomes produced a cost–benefits result for the ITS scenario that was significantly higher than the cost–benefits result obtained for the "no-change" scenario. This demonstrated that implementation of an ITS was, given information available at the time, a cost-beneficial strategy.

The combined information provided by the technology and the cost–benefits analysis set the stage for client buy-in to the IT project.

DEVELOPING ADVANCED TECHNOLOGY TRAINING APPLICATIONS

The development of LEAP followed an iterative design process as espoused by Buxton and Schneiderman (1980) that involves rapid prototyping to get feedback from potential users as early as possible. Each iteration of the system involves first summarizing feedback from the current implementation and then extending the software. Within each iteration, we employed many of the activities that comprise the usability engineering lifecycle (Nielsen, 1993), including knowledge of the user, participatory design, iterative prototyping, and heuristic and formal evaluations. The development has gone through three main iterations so far.

Part of the reason for a "phased" or iterative development process had to do with the way industry conducts research and development on high-risk technologies. Generally, high-risk efforts are costly, and time and labor intensive. Couple that with the fact that only a low percentage of these efforts ever make it to deployment, and it is not surprising that industry employs a stage-gate process with go/no-go decision points. In this section, we first describe the most recent version of LEAP. For a more complete description, readers are referred to Bloom et al. (in press). Following that, we describe each of the three development phases.

The LEAP System Overview

LEAP is an ITS capable of intelligently coaching users on how to perform the job of a customer contact employee (CCE). As described previously, the job of a CCE is to be a service company's primary interface to its customers. To do this, a CCE must be able to simultaneously: (a) converse with customers to solve problems and sell services, (b) manipulate phone calls, (c) correctly enter data into service registration, repair orders, and billing software while maneuvering through numerous screens and fields, (d) understand all features and incompatibilities of offered services, and (e) look up information about service availability and capabilities from reference documentation.

LEAP is a general, multimedia ITS platform that emulates the CCE's job environment by providing simulated customers, expert CCEs, and CCE software applications. Within this simulated environment, LEAP provides realistic learning activities in the form of scenarios to practice; a coach to assist users in working through those scenarios by providing assistance as needed or requested as well as applying user-appropriate, proactive instructional strategies; and an opportunity to review their performance at the completion of each scenario.

In LEAP's simulated customer contacts, users proceed through a series of interactions, either dialogue interactions with LEAP's simulated customers or application interactions with LEAP's simulated software applications. Some examples of conversational actions taken by a CCE are: answering the phone with the name of the company, asking the customer for his or her phone number, or telling the customer whether a particular service is available. Examples of application actions are: entering the customer's phone number into a particular field on an application screen, navigating to a new application screen to locate information, or concluding from information displayed in an application that a particular service is available where the customer lives.

As depicted in Fig. 4.1, the LEAP Practice Environment consists of the LEAP Practice Commands window (top left), the LEAP Practice Conversation window (bottom left), one or more Application Simulator windows (top right), and a Contact History window (bottom right).

At each step in a customer contact, LEAP allows the user either to observe how an expert CCE might respond to the current situation or to practice handling that same situation himself or herself. This is accomplished by employing three instructional modes: Observe, Practice, or Focused Practice.

In Observe mode and Practice mode, each action in the current situation is either observed or practiced, respectively. In Focused Practice mode, LEAP uses its student model to determine whether it is most beneficial for the user to observe or to practice the actions involved in the current situation. The mode for conversation actions is set independently of the mode for application actions. For example, if the current stage of the conversation calls for the CCE to look up in a database whether a particular service is available in the customer's area and inform the customer of this,

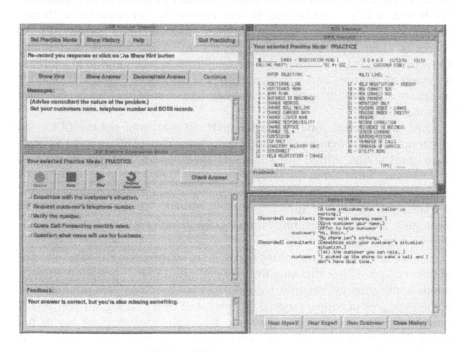

FIG. 4.1. LEAP practice environment.

LEAP might determine that the user should observe how to locate this information in the application but practice telling the customer whether the service is available.

When in Observe mode, or when in Focused Practice mode and LEAP has chosen to have the user observe conversational actions for the current step in the contact, the interaction proceeds in the following way: First, the user hears a recording of what the customer says. After this, a list of appropriate actions to take is displayed for the user. The user may review these actions and then press a Demonstrate Answer button, which plays a recording of an expert CCE responding to the customer with the listed actions. This expert recording models for the user an appropriate response in the current context, both in terms of what is said and how it is said.

When in practice mode or when in Focused Practice mode and LEAP has chosen to have the user practice the current step, the user first hears the customer recording and then is prompted to record an appropriate response using the recorder panel in the Practice Conversation window (Fig. 4.2).

After the user records a response, a list of possible responses is displayed below the recorder panel. The user selects the combination of answers that

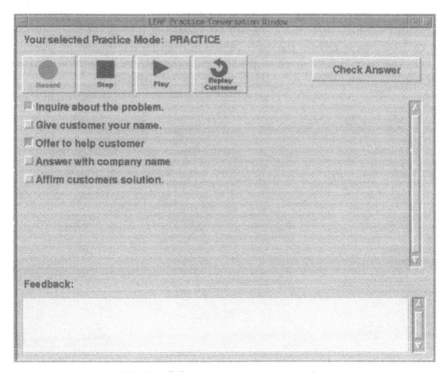

FIG. 4.2. LEAP practice conversation window.

best describes his or her recorded answer and then clicks on the Check Answer button to get feedback on the selected responses. The feedback is displayed textually at the bottom of the Practice Conversation window. If the user's response is appropriate, he or she may continue to the next step in the contact by pressing the Continue button in the LEAP Practice Commands window (Fig. 4.3). Before continuing, the user may also elect to hear a recording of an expert responding to the current situation by pressing the Demonstrate Answer button in the Practice Commands window.

If the user's selected responses are not appropriate, textual feedback to this effect is displayed. The user may then try different combinations of answers, getting feedback for each. The user can also rerecord his or her response. At any time, the user can go to the LEAP commands window and ask for a hint, or, after seeing a hint, the user may ask to have the answer shown or demonstrated by the playing of the expert recording. Because there may be more than one appropriate way of continuing the contact from any given point, allowing users to try different combinations of answers and get feedback on each provides a flexible means of exploring the space of appropriate ways of responding to the current situation.

When either the user or LEAP has elected to practice the current application actions, the user may then enter information into the application, navigate to other screens to find information, and so on. Each application simulator window includes a display for feedback on the appropriateness of each action taken by the user. When the applications are in Observe mode, the user is prompted to press the Demonstrate Answer button for each application action to be taken in the current situation.

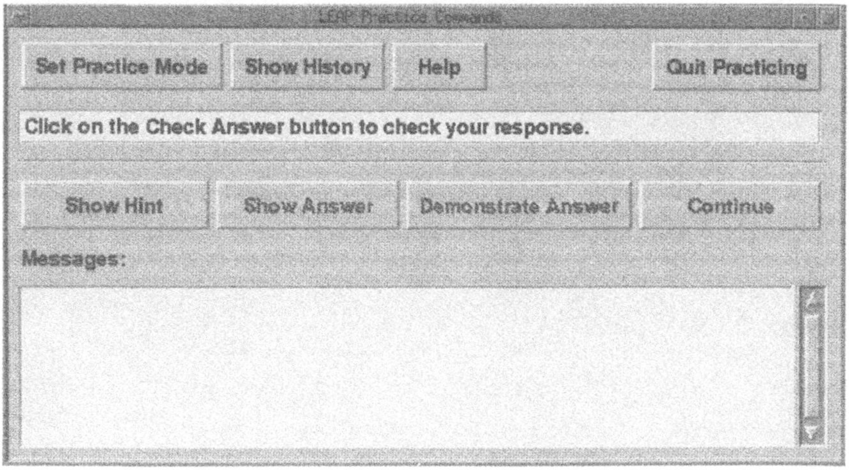

FIG. 4.3. LEAP commands window.

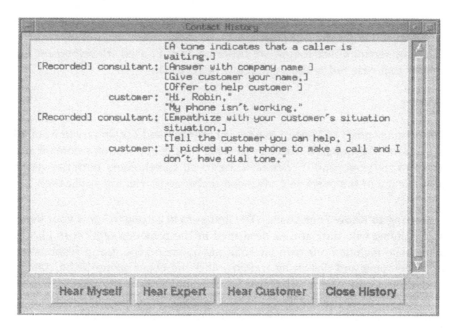

FIG. 4.4. Contact history window.

The Practice Commands window also contains a prompting area that indicates what the user needs to do at any point in the contact. In addition, the practice mode may be set by clicking on the Set Practice Mode button that invokes the Set Practice Model dialogue box. Users can also get "help" with how to proceed by clicking on the Help button. The Show History button controls the display of the Contact History window (Fig. 4.4). This window contains a scrollable list of the interaction between the customer and the CCE. The user can review any item by selecting it and then choosing to hear recordings of the customer, the expert, or himself or herself.

Any time users exit a contact, either prematurely or on completion of the contact, they are given the option of going to a Reflective Follow-up environment before selecting another contact to practice. In Reflective Follow-up, there are three tools. The first tool, "Performance Review," enables users to examine LEAP's assessment of their current skill level and to observe both short-term and long-term changes and trends in their skill levels. The second tool, "Self Critique," enables users to assess their own performance on the contact just practiced on a number of subjective areas that LEAP cannot measure but are considered by content authors to be important to contact expertise. The third tool, "Multimedia Review," enables users to view any of a set of selected multimedia summaries describing either salient points about the contacts practiced or errors that novice CCE's commonly make, and their consequences. In addition, for each summary area, users can access that part of the contact just practiced that is related to the topic for review and playback.

In the sections that follow, we describe the "Phased" development approach that resulted in the system just described.

Phase I

The primary goals of Phase 1 are to: (a) identify and obtain buy-in from all "stakeholder" groups, (b) reach consensus on the application's domain and scope of coverage, and (c) demonstrate to all stakeholders both the vision and efficacy of the proposed advanced technology training application.

Getting to Know Your Users. The first step in getting to know your users is identifying who they are. As described in the previous section, in LEAP's case these included our own internal management, the group responsible for developing and delivering training, and the client group to be trained (and their management). However, at this point, it is insufficient to just identify impacted organizations—at this point, it is necessary for members of those organizations to come forward and actively participate in and/or sponsor the effort.

In the LEAP project, our method for dealing with this issue was to arrange a kick-off meeting with "formal" representatives from all stakeholder organizations. The kick-off meeting combined presentations of the technology and its applicability with working sessions to obtain consensus on the requirements and scope of the system to be developed, including identifying resources to be provided by each organization and project roles and tasks to be performed by *all* participants.

Participatory Design. Now that all stakeholders have been identified and roles and responsibilities agreed on, the next step is to gain an understanding of the tasks your end users perform and how best to provide them with training for those tasks.

As a starting point for understanding CCEs and their job, we initiated two activities. The first activity involved working with a consultant affiliated with the learning organization who had several years of experience developing training for CCEs. To help familiarize the development team with the domain, the consultant began by drawing diagrams describing the branches that can happen in typical CCEs' dialogues with customers. As it turned out, these diagrams had a direct mapping to Recursive Transition Networks (RTN) that are simplified Augmented Transition Networks (ATN) most commonly used in natural language processing (Woods, 1970).

This RTN representation was adopted for the domain expertise component of LEAP for several reasons. First, the representation allows actions and combinations of actions that are common among more than one conversation to have a single definition. By having conversations integrated into a single representation, a tutor could easily use knowledge from a student's performance on one conversation to better teach future conversations. Second, because the RTNs represent a generalized description of a set of conversations, an RTN may be used to automatically generate variations on the conversations used as a seed for the RTN. Third, if domain experts conceptualize dialogues in an RTN-like model, then making that LEAP's representation might facilitate both subject-matter expert authoring of LEAP knowledge bases as well as the development of accurate and appropriate conceptual models in trainees.

The second activity involved the collection of actual customer contacts for the domain identified from end users out in the field. Not only did this provide us with excellent input to help in the development of the knowledge base but it also served to get members of the CCE end-user population both informed and involved in our efforts, something that eventually paid tremendous dividends later in the project.

Armed with this information, our next task was to design the ITS architecture that supported the delivery of ITS-based training in customer contacts. First, we wanted to develop an accurate, yet manageable emulation of the domain along with a means of supporting the users' interactions with that emulation. Users needed to "hear" customers and respond to them verbally (as is done on the job). However, given limitations in natural language understanding, we needed an alternate mechanism to enable the users to communicate to the system (as unobtrusively as possible) what they had just said. In addition, we needed to provide users with emulations of the software systems they interact with in the normal course of their job.

Second, we wanted to apply an actual theory of instruction in LEAP. To accomplish this, we based LEAP's instructional model on two existing apprenticeship models: (a) Collins, Brown, and Newman's (1989) "Characteristics of Ideal Learning Environments," and (b) the intelligent-coached apprenticeship framework of Sherlock (Lesgold, Eggan, Katz, & Rao, 1992; Lesgold, Lajoie, Bunzo, & Eggan, 1992). In addition, we incorporated several features of the minimalist approach to learning (Carroll, 1990) in LEAP:

• Teach people what they need to learn in order to perform their jobs. Training should take place in an environment that closely emulates their real-world job environment, and the tasks they perform are representative of tasks they will perform on the job.

• Allow learners to start immediately on meaningful and realistic job tasks, and allow them to work on those tasks in any order. Instruction should be user initiated. Even in the case where the ITS has the capability to recommend instruction, the user should have the choice of whether to take that recommendation.

• Keep the amount of passive instruction (i.e., reading) to a minimum. All learning activities should center around performing a job, not learning about the job.

Third, we had to develop a method of tracking and assessing student knowledge that allowed us to apply this instructional model on an individual basis. LEAP's student model is based on the observation that the probability of a correct response increases as a function of the number of previous correct responses (Newell & Rosenbloom, 1981; Underwood, 1954). User performance data is collected and stored in the student model at a fine-grained level, that of individual user actions. Action skill-level values are used to determine whether, given a choice, LEAP should model the performance of the action for a user, allow the user to quickly see the action being performed if it has already been mastered, or ask the user to attempt the action himself or herself.

Finally, we wanted to make sure that the architecture and system we developed were "generic," that is, they could be used to cost effectively deliver any training application that fit the CCE model: simultaneously conversing with customers and manipulating any number of applicable software applications.

To make an ITS platform generic and cost effective requires being able to reuse as much of the architecture and actual code as possible. In an ITS, it is a given that the knowledge base (or domain content) must be crafted anew for each training application. However, at a minimum, to be generic means being able to reuse student modeling, instructional modeling, and knowledge-base inferencing mechanisms from one training application to the next. In addition, to be cost effective requires an economy in the use and construction (or reconstruction) of the domain emulation and user interface of the ITS. Something we had not anticipated was that not only was the user interface (including the application simulations) completely reusable from one training application to the next but the domain emulation did not require major construction or reconstruction efforts from one training application to the next, because the auditory dialogue portion of the emulation was embedded in the knowledge base. This made LEAP an extremely cost-effective platform. Therefore, by constructing the knowledge base (a necessary effort), one also crafts the domain emulation for that knowledge base.

Given these design constraints, our next step was to develop in parallel a sample knowledge base of dialogues for the specific training domain chosen and the actual ITS used to deliver that training.

Rapid Prototyping. The knowledge base was developed using a traditional knowledge-engineering approach. Development using this methodology was slow and time consuming due to problems such as knowledge filtration, consensus building problems, and so on, as observed in the knowledge acquisition literature (Bloom, Bullemer, Nasiruddin, Ray, & Penner, 1992). System development proceeded rapidly. However, speed was gained at the cost of important software engineering disciplines designed to ensure robustness, such as documentation, code comments, code tests, and so forth.

The end result was a working system that demonstrated many important features of an ITS, such as guiding students in exploratory learning, adapting lesson plans to the individual students, and providing useful feedback during problem solving. Specifically, the prototype permitted students to observe or practice a conversation, record and playback their responses, and use the order entry software. Although this version of LEAP showed progress to our clients and demonstrated some of the power of ITSs, this version did not have a real coaching capability; it was not sufficiently robust to stand real students beating on it, and course material was not fully developed and verified with the clients. However, the version did suffice to gain stakeholder approval to continue our efforts on to the next phase—the development of a robust, field-testable prototype.

Phase 2

We had now succeeded in demonstrating to stakeholders both the vision and efficacy of advanced technology training applications, gaining approval to continue development of the system in preparation for a comprehensive evaluation in the field. However, along with this approval came some conditions.

Dealing With "Preflight Jitters." Now that client representatives had decided that the system was going to be used to train actual prospective CCEs in the operational environment, even for the purpose of evaluating its efficacy, they wanted tighter control over what was to be trained. In Phase 1, it did not matter if there were some inaccuracies in the knowledge base as it was only used for demonstration purposes. However, now faced with the prospect of actual CCE trainees using the system, the clients wanted tighter "checks" on the knowledge base, including formal sign-off for both content (the topics and specific "teaching points" of each contact) and language (the specific words spoken by LEAP's expert CCE—thus modeled for the trainee). This is not to imply that there were major inaccuracies in what had been built up to that point. In fact, management feedback to that effect was relatively minor. Rather, the constraints reflected their increased anxiety due to the risky nature of the technology involved.

In effect, the client representatives were imposing requirements on the project that would never be imposed on traditional training products. In a traditional CCE class, instructors create role plays on the fly, often including inaccurate information or "bad" language in their scenarios. However, no one insists they need to check their content or language even though they are not as well thought out as the scenarios included in LEAP. This was particularly frustrating when one considers that the subject-matter experts (or SMEs) providing the input for the construction of the LEAP contacts represent the same population of people who develop the traditional training (and their role plays).

In response to these concerns and to further our client's positive perceptions about the project and the development team, we instituted the following measures. First, we increased the scope of SME participation by moving to a team of experts (Cochran, Bloom, & Bullemer, 1990) that included CCEs, instructors, and instructional designers selected by our client representatives. Second, as the CCE job is a union job, we solicited direct involvement and approval from union representatives by having periodic meetings with them and insisting the CCE representative on the SME team also represent the union. Finally, we instituted a formal approval process for both knowledge-base content and language. In the approval process, a team of "management" experts was identified by our client organizations. These management experts were responsible for reviewing the content and language of the contacts in a timely manner, providing the SME team with signed documentation indicating either approval, approval with change recommendations, or rejection with explanation.

These activities seemed to allay most of the anxiety that clients had with the content of the system, but we still had to address their concerns regarding the technology itself. In particular, clients wanted some assurance that the system's "artificial intelligence" was able to accurately model students' skills and knowledge and was able to use those models to deliver appropriate, individualized instruction. In addition, they felt the user interface employed in the Phase 1 version was too complex.

To allay clients' anxieties concerning the risks of ITS technology, we conducted comprehensive formative and summative evaluations of LEAP (Bloom, Linton, & Bell, 1993). In the formative evaluation, the objective was to evaluate the LEAP architecture and its behavior, seeking to answer the questions: "Does LEAP work as expected?" and "Can CCE trainees use LEAP effectively?" The conduct of this evaluation involved the application of specific evaluation techniques (i.e., expert inspections of the knowledge base, sensitivity analyses of the student model, and pilot testing of the entire system) within the context of our iterative, participatory design process. In the summative evaluation, the objective was to evaluate LEAP's instructional and emotional impact. The summative evaluation was conducted with five CCE initial training classes, seeking to answer the questions: "Is LEAP an effective learning tool?" "Do trainees like using LEAP?" and "Is LEAP easy to use?" Data collection was from a number of questionnaires (i.e., measures of sentiment) and LEAP usage logs and student models for each trainee.

The results of this evaluation indicated that, in general, both trainees and instructors liked using LEAP, found it easy to use, and felt they learned a lot using LEAP, a finding validated by LEAP's usage logs and student assessments (Bloom, Linton & Bell, 1993). However, the evaluation also uncovered limitations with the Phase 2 version and provided input into Phase 3 development activities.

Phase 3

The goal of Phase 3 is to produce a robust, fully tested and documented version of the advanced technology training application. In most software development projects, this phase is considered straightforward. However, in the case of an advanced, interactive learning environment application, nothing is ever straightforward.

Adapting to Change. The most significant event that occurred between Phases 1 and 2 was the initiation by our company of re-engineering efforts that resulted in a redesign of its very business structure. In other words, the client organizations we had worked so hard to educate and involve were no more. We were cast into the situation of having to identify who our new clients were and to educate and socialize them to the technology, as we had done with our original clients.

Fortunately, we were aided in these efforts by the client group for whom the system was developed—the CCEs. LEAP had developed a formidable "underground" reputation among the CCE community based on our participatory design and evaluation efforts with the CCE population. When these same CCEs were asked by the management of their "new" organization what could be done to help them in their jobs, one of the suggestions brought forward was to provide them with LEAP training. This initiated a chain of several months of demonstrations and presentations that eventually culminated with a decision by the new client organization to go forward with a deployment of LEAP.

Given our new clients and the directions they saw regarding the use of our system, Phase 3 introduced additional constraints. First, our new clients' largest CCE training concern was facilitating the transition from sales-oriented customer contacts (where the goal is to sell as many services to the customers as possible) to "consultative" contacts (where the goal is to help the customers identify their needs and solve their problems). It was determined that we should produce a LEAP training module for this need. Second, to demonstrate the generality and cost effectiveness of the platform, we needed to have this new training module developed entirely by a SME team without formal-knowledge engineering support. Third, from a system perspective, we needed to address the system's "shortcomings" identified in the Phase 2 evaluation. Fourth, our impending technology transfer required that the system be fully tested and documented.

To truly demonstrate to our new clients that ITSs can be a cost-effective training tool, we had to determine if it was possible for nontechnical, subject-matter experts to develop a LEAP training module from *scratch* and with minimal support from the development team. This task was further complicated by the fact that we had only rudimentary tools to support their efforts—tools not of the normal quality provided to end-user/authors. Our primary reason for doing this was to give us the opportunity to observe the entire authoring process to help us discover what skills LEAP authors needed and what types of tools best support the authoring effort. Chapter 5 of this volume by Anne McClard describes this authoring process study and findings.

In general, the knowledge base produced by the authoring team far exceeded our expectations. Even though they initially had some difficulty understanding LEAP's abstract knowledge representation scheme (something that was not surprising, given their lack of technical backgrounds), they managed to develop a training module and knowledge base whose complexity and sophistication were far beyond anything we might have imagined (or, for that matter, developed ourselves). This demonstrated to our clients that ITS authoring could be done by subject-matter experts (given a reasonable learning curve for understanding LEAP's representation scheme) and that it could be just as cost effective as CBT authoring.

With regard to the LEAP software itself, the Phase 2 evaluation had identified three major limitations in the platform. First, in that version, contacts, as represented, lacked flexibility in the way in which they were carried out. In each contact, LEAP's representation only allowed one way of performing the contact (that is, of sequencing the activities in that contact). Analyses of more experienced CCEs (who would likely be the end-user population of this latest version of LEAP) indicated that there was a great deal of flexibility in the way contacts were actually carried out, both in the ordering of actions and in the ability to connect different combinations of actions at a single stage of the contact to satisfy multiple objectives. In the Phase 2 version, the ordering of actions was predetermined, and LEAP only supported the satisfaction of a single objective at any state of the contact. In the Phase 3 version, we made changes in the system to support these more flexible types of contacts observed in more experienced CCEs.

Second, in the Phase 2 version, LEAP provided only very limited opportunities at the completion of a contact to review and reflect on what they had just practiced. Research has found that reflection enables users to identify areas where additional learning work needs to be done, generalize their knowledge from specific examples to general cases, and elaborate on their conceptual knowledge, particularly in the area of explanations of expert behavior or processes (Bielaczyc, Pirolli, & Brown, 1994a, 1994b; Chi, Bassok, Lewis, Reimann, & Glaser, 1989; Chi & VanLehn, 1991; VanLehn, Jones, & Chi, 1992). In the Phase 3 version, we enhanced LEAP to support greater amounts and types of reflection on exiting a contact rehearsal (as described in the previous section containing a brief description of LEAP).

Finally, given the feedback from both the Phase 2 evaluation and our own observations of the complexity of the LEAP user interface, we decided to initiate a comprehensive usability evaluation of LEAP. The end results of this evaluation were that we were able to consistently apply a user interaction metaphor for LEAP that complied with existing user interface guidelines and standards while significantly reducing the complexity of trainees' interactions with LEAP.

With regard to readying the system for eventual technology transfer, with an interactive learning environment application it is necessary to perform additional software engineering lifecycle work. In the LEAP project, this consisted of three tasks. First, we ported all code to a single platform, made sure the code was well commented, and had design documents that supported a new developer/maintainer learning about the system in the shortest time possible. Second, we engaged the services of a formal software testing organization. Third, we produced a user guide to support the system's use in the field and an installation and maintenance guide for the operational support organizations.

At the completion of the third phase, we had produced a working system that is currently being used in our first field deployment (Bloom, Bell, Meiskey, Sparks, Dooley, & Linton, in press). This version improved on a number of areas such as dealing with multiple solution paths from a student, providing an environment for students to reflect on their problem solutions, and presenting a richer problem-solving environment. However, successfully deploying an advanced interactive learning environment involves much more than the production of a "deployable" system. In the next section, we discuss many of the issues that impact successful deployment and how we addressed those issues in the LEAP project.

DEPLOYING ADVANCED TECHNOLOGY TRAINING APPLICATIONS

To deploy an advanced interactive learning application, an ITS, in industry, considerable attention must be paid to three major issues: (a) educating decision makers about the need to re-engineer how training is conducted and about the power of an interactive learning environment (ILE), (b) promoting or socializing ILE technology to all impacted user groups, and (c) working "within the system" to ensure the ILE meets all deployment requirements for the company and/or organization. Failure to pay serious attention to any of these issues could be fatal to the technology transfer effort. In the sections that follow, we discuss each of these issues in detail, including how they were addressed in the LEAP project and what the eventual outcome of those efforts was.

Educating Decision Makers on the Need to Reengineer

The fundamental mistake most companies make when they look at advanced technology applications is to view them in terms of their existing processes (Hammer & Champy, 1993). Most corporate decision makers attempt to cast technology advances such as interactive, computer-based learning environments in terms of how they can improve what they are already doing. However, the real benefit of technology is to support attainment of entirely new goals; to think how technology can enable us to do things that we cannot now do (Hammer & Champy, 1993).

As Hammer and Champy (1993) pointed out, most corporate decision makers know how to think deductively: to define problems and seek out and evaluate different solutions to that problem. Technology demands inductive thinking—the ability to realize a powerful solution and then seek problems to solve with that solution the company may not even be aware it has. Therefore, it is of no use to ask people how to use technology in their business as they will inevitably reply in terms of how that technology might improve a task they already do. In addition, most corporate decision makers may view technology as disruptive in that it breaks well-established rules on how to do one's work (Hammer & Champy, 1993). However, it is this disruptive power of technology that can help promote inductive thinking, producing major paradigm shifts in the ways we do business.

Hammer and Champy (1993) proposed that to think inductively about the uses of technology, one must look at the "rules" about work and uncover ways in which technology can break those rules, resulting in new rules and new ways of doing business. In the area of training, the old rule is learning is taking in information (Senge, 1990). This traditional view of learning has its origins in the schoolhouse model of learning and education (Collins, Brown, & Newman, 1989). However, taking in information is only distantly related to real learning. Real learning involves learning from experience—performing activities, making decisions, and directly experiencing the consequences of those activities and decisions (Collins et al., 1989; Lave & Wenger, 1991; Senge, 1990). However, experiential learning is only effective if the feedback one receives in response to one's actions and decisions is rapid and consistent with reality.

Senge (1990) advocated the use of "microworlds"—specifically designed microcosms of reality within which people can learn through experimentation with, or operations on, the objects in that microworld (Papert, 1980). In terms of technological advances, interactive learning environments (such as ITSs) go well beyond the notion of a microworld. A central component of an ITS is a simulation of the real-world task environment (or microworld). However, ITSs add to their microworlds the ability to guide learning activities, both reactively and proactively, as well as the ability to provide learners with ongoing assessments of the skills.

The process to be employed by technologists attempting to convince corporate decision makers about the efficacy and utility of ILEs is as follows: (a) Help them to think inductively. Present the technology in such a way as to support understanding of how that technology "breaks the rules," resulting in a reengineered and more effective training; (b) educate them on the value of experiential learning. Allow them to experience its power and see firsthand the need to reengineer their training model; (c) demonstrate how ILEs extend this powerful notion of microworlds. Present the logical argument that if microworlds are desirable, then technologies that extend that notion are even more valuable.

In retrospect, most of our efforts to gain corporate acceptance for LEAP followed this process more as a function of fortunate circumstance than as the product of good planning. However, we were able to realize that if not for our good fortune, we might have gone the way of many good research projects—nowhere. This has led us now to be prescriptive in the application of this process.

In the LEAP project, we were fortunate to have had the support of a set of forward-thinking managers from our client organization—managers who listened to their employees when they complained that the training they were receiving was inadequate and that they needed training that better prepared them to perform their job. In addition, through our participatory design and development efforts, we cultivated an extremely loyal following within the end-user population. These two circumstances combined to bring LEAP to the attention of powerful decision makers currently leading the efforts to reengineer part of the company's business processes. Given their predisposition toward the role of technology in reengineering, we were able to gain their sponsorship. However, the sponsorship only served to "unlock the door"; to succeed, we still had to prove our worth to all impacted users. Such influential sponsorship also heightened expectations regarding the quality of the to-be-delivered ILE. In the next two sections, we discuss these two issues.

Promoting and Socializing Advanced Technology Training

We now had the sponsorship of the corporate decision makers. However, we still had to deal with the populations of impacted users (i.e., workers to be trained using the ILE, instructors who must incorporate the use of the ILE in their training plans, and training developers who must use the ILE to craft meaningful learning activities) who were certain to be both unsure and suspicious of this technology, manifested by a "resistance to change."

Resistance comes from threats to traditional ways of doing things, which themselves can be deeply rooted in established power relationships (Senge, 1990). Workers to be trained by the ILE will most likely resist its use through cynicism. They have been promised many things in the past and will view the ILE as just another opportunity for the company to "disappoint" them. Instructors who must use the ILE as part of their training program and training developers who must craft meaningful learning activities into the ILE will view it as a threat, both to their established way of operating and to their job security.

The answer to dealing with this resistance is not to use force as this will only increase the strength of the resistance. The answer involves a combination of education and socialization (i.e., making people feel comfortable about the technology and giving them a sense of ownership). Just as corporate decision makers need to be educated on the efficacy of ILE in re-engineering training, so do impacted users.

Workers need signs from management that there is a real commitment to this change. This may be the single most imposing obstacle to re-engineering efforts. It requires that all employees at all levels have a shared vision or model about the re-engineered process, the role ILEs play in that re-engineered process, and the impact that role has on the rest of their jobs. For example, it is not good encouraging workers to take the time to hone their skills in an ILE if the time spent using that ILE (and away from their job) impacts their personal revenue.

Instructors and training developers need to be educated (and convinced) about the benefits of ILEs and of the need to reengineer how training is done. In addition (and possibly more importantly), they need to be reassured that the ILE and the changes it brings about are not a threat to their job status or security but that the ILE is a tool that when applied effectively, should enhance their job status and security.

Unfortunately, we had no control over ensuring that all parties (i.e., workers and management) shared the same vision for the re-engineering of training. In fact, this is still an open issue. The best we could hope for was that both parties trusted in our sincerity and desire to succeed. At present, we have achieved modest success in receiving buy-in from workers, instructors, and training developers who were in some way connected to the project efforts. What was and still is a problem is that some of the layers of management from the different organizations involved did not share their leaders' vision, resulting in considerable disagreement over the deployment of LEAP and the reengineering of training in general. This is not an uncommon problem encountered in re-engineering efforts. Allowing existing corporate cultures (such as "this is the way we have always done training") and management attitudes to become obstacles has often been the downfall of re-engineering efforts. However, the answer is not to give up but to continue to try to educate, an effort we are pursuing.

Working Toward Deployment

Another issue to be addressed, one no less important that the previous two with regard to deploying advanced technology training applications, is working within the system to ensure the application meets all deployment requirements for the company or organization. Deployment requirements include: conducting system tests, preparing user documentation, producing detailed project plans and reports, working with procurement organizations, and working with operations support groups. Failure to attend to these requirements can adversely impact the application's acceptance by those organizations, a factor critical to successful deployment.

What complicates deployment of an advanced technology training application such as an ITS is that it is far more complex than the types of systems deployment that personnel normally encounter. ITSs are special purpose systems that require that deployment personnel have a background in object-oriented design, programming and languages, artificial intelligence, and human learning. However, even having a sufficient background in these areas does not guarantee they will have the same understanding of the way the system is supposed to work as the system's designers and implementers. Given the complexity of an ITS, it requires extraordinary measures to ensure all parties responsible for both deployment and maintenance have a sufficiently accurate understanding of the system that they do not introduce errors by violating its basic design and purpose.

These extraordinary measures include: (a) integrating deployment personnel on the project with development personnel as early as possible, (b) providing opportunities to educate deployment personnel on the system's basic design and purpose, and (c) producing software documentation that not only captures system requirements and design but that also captures information on the way things are supposed to work, often maintained informally among developers as part of the "folklore" of the system.[3]

---

[3]These ideas have been proposed by Greg Clemenson, software engineering consultant to the LEAP project.

Integrating deployment personnel into the project as early as possible is far from a simple matter. Deployment personnel and organizations are used to working with familiar systems. This has created an approach whereby they anticipate that a very simple transfer can be accomplished in short order and at the very last moment. However, this model does not leave them with much time or resources to dedicate to a long-term technology transfer effort in which they must learn about a new type of system. This is an ongoing problem and not one likely to be solved without a demonstration of the need to re-engineer the system. Developers of advanced technology applications can only hope that as more and more of these applications are deployed, managers and members of deployment organizations will see the need to re-engineer the technology transfer process and to become "partners" with developers.

It should come as no surprise that the theme of educating one's clients, users, and deployment personnel occurs repeatedly throughout this chapter, given that is the area in which we have chosen to work. However, it does not just reflect a bias in our belief structures that education is good and everyone needs more; it is also an effective way of dealing with these problems. This does not mean one should hold classes and teach about the system in the traditional sense, but, rather, the primary purpose for the integration of deployment and development personnel is to create an apprenticeship environment where deployment personnel will have the opportunity to learn about the system's basic design and purpose in the context of helping develop and maintain the system (Collins, Brown, & Newman, 1989; Lave & Wenger, 1991).

Unfortunately, as stated previously, it is not always possible for deployment personnel to spend large amounts of time with development personnel so that apprenticeship learning can take place. In such cases, one must take other measures to facilitate learning. On the LEAP project, we were fortunate to have the software engineering support of a consultant experienced in the transfer of advanced technology applications. This consultant proposed a new approach to software documentation that enhances existing documentation about system requirements and design to also capture information on the way the system is supposed to work. The method the consultant proposed for capturing the way a system is supposed to work is through "scenarios." Scenarios are documents that tell the reader about what is supposed to happen under certain circumstances. Scenarios should provide deployment personnel with a simple way of judging whether the program is operating as expected under particular situations.

Another type of knowledge to be transferred has to do with how to fix and extend the system using an environment of utilities and tools crafted by the development team. These tools were specifically crafted by the developers to help them maintain and debug problems in the code. In conjunction with these tools and utilities, developers can create a second class of scenarios documented to describe common problems encountered with the code and how those problems were or could be addressed using the existing utilities and tools.

CONCLUSIONS

We believe that our experiences with LEAP have given us insights into what it takes to successfully initiate, develop, and deploy advanced, interactive, learning environment applications. We are the first to acknowledge that not everything we did turned out to be necessary, or was done correctly on the first try. Much of this can be attributed to the fact that we were learning as we went along. In addition, it is difficult to define a single, definitive process that applies to all advanced technology development efforts, as the stages of any process (and the activities within those stages) are contextually dependent on the specifics of each situation. However, in retrospect, we can identify several principles we feel can facilitate such efforts.

First, cost–benefits analyses can be useful in helping management and client organizations decide whether to undertake the development of an advanced technology application. In our case, a cost–benefits analysis with projected business impact was a critical factor in the process of deciding whether to fund LEAP.

Second, it is important to educate and involve both management and end-user representatives of all user groups impacted by the application early in the effort and to maintain their involvement throughout the life the effort. One day, the system you are developing will cease to be "yours" and instead will be "theirs." Preparing for and encouraging this eventuality, even from the onset of the project, can help your development and deployment efforts.

Third, plan for how deployment and technology transfer will happen from the research and development organization to the operational unit right from the start. Maintaining an advanced technology application requires different skills than maintaining "traditional" software applications. Organizations that will eventually assume responsibility for these applications need to undergo changes in both their general skill sets and in how they operate. In addition, because such education and change take time, the best way of facilitating their happening is to integrate support groups with development groups as early as possible.

Fourth, employ a phased-development approach with go/no-go decision points between phases. Such a stage-gate approach has several benefits. First, it helps in the management of client expectations. Having clearly defined goals for each stage not only supports progressive development (i.e., functional prototype, field prototype, deployment system) but also keeps clients from expecting a finished product before completion. Second, a stage-gate approach helps control resource allocation. Although we hate to acknowledge this, sometimes our ideas do not work out as expected or desired. A phased approach allows for early evaluation of application feasibility without having to commit the bulk of your resources.

Finally, be flexible with your clients and with your own expectations. Businesses change, more often and more rapidly than expected. In the course of a multiphase, advanced, technology-application development effort, some aspects of your organization are bound to change. You need to adapt to these changes and face them positively, even if it means duplicating efforts conducted earlier in the project. Failing to adapt is the surest way to bring a premature end to your project.

REFERENCES

Aronson, D. T., & Briggs, L. J. (1983). Contributions of Gagné and Briggs to a prescriptive model of instruction. In C. M. Reigeluth (Ed.), *Instructional design theories and models: An overview of their current status* (pp. 75–100). Hillsdale, NJ: Lawrence Erlbaum Associates.

Bielaczyc, K., Pirolli, P., & Brown, A. (1994a). *Cognitive explanations and metacognition: Identifying successful learning activities in the acquisition of cognitive skills* (Report No., CSM-8). Berkeley, CA: University of California at Berkeley.

Bielaczyc, K., Pirolli, P., & Brown, A. (1994b). *Training is self-explanation and self-regulation strategies: Investigating the effects of knowledge acquisition activities on problem-solving* (Report No. CSM-7). Berkeley, CA: University of California at Berkeley.

Bloom, C. P., Bell, B., & Linton, F. (1994). The learn, explore, and practice intelligent tutoring systems platform. In M. M. Tanik, W. Rossak, & D. E. Cooke (Eds.), *Software systems in engineering* (pp. 357–366). New York: ASME Press.

Bloom, C. P., Bell, B., Meiskey, L., Sparks, R., Dooley, S., & Linton, F. (1995). Putting intelligent tutoring systems technology into practice: A study in technology extension and transfer. *Machine Mediated Learning, 5,* 13–41.

Bloom, C. P., Bullemer, P. T., Nasiruddin, M., Ray, J., & Penner, R. (1992). *Artificial intelligence in training (AIT).* (Research Report No. AL-TP-1992-0010). Brooks Air Force Base, TX: Air Force Human Resources Directorate, Technical Training Research Division.

Bloom, C. P., Linton, F., & Bell, B. (1993). *An evaluation of the learn, explore, and practice intelligent tutoring systems platform.* (Research Rep. No. AT-09_10-002817-00.01). Boulder, CO: US WEST Technologies.

Buxton, W. and Schneiderman, B. (1980). *Iteration and the design of the human computer interface.* Proceedings of the 11th annual meeting of the Human Factors Association of Canada, 72–81.

Carroll, J. M. (1990). *The Nuremberg funnel.* Cambridge, MA: MIT Press.

Chi, M., Bassok, M., Lewis, M., Reimann, P., & Glaser, R. (1989). Self-explanations: How students study and use examples in learning to solve problems. *Cognitive Science, 13,* 145–182.

Chi, M., & VanLehn, K. (1991). The content of physics explanations. *Journal of the Learning Sciences, 1,* 69–105.

Cochran, E. L., Bloom, C. P., & Bullemer, P. T. (1990). Increasing end-user acceptance of expert systems by using multiple experts: Case studies in knowledge acquisition. In K. McGraw &

C. Westphal (Eds.), *Readings in knowledge acquisition: Current practices and trends* (pp. 73–89). New York: Ellis Horwood.

Collins, A., Brown, J. S., & Newman, S. E. (1989). Cognitive apprenticeship: Teaching the crafts of reading, writing, and mathematics. In L. B. Resnick (Ed.), *Knowing, learning, and instruction: Essays in honor of Robert Glaser* (pp. 453–494). Hillsdale, NJ: Lawrence Erlbaum Associates.

Computer-Based Learning Will Benefit From Sharp Budget Increases, *HRFOCUS* (July 1994), 22.

Coping with a skills shortage, *Institutional Investor* (30 July 1994), 29–30.

DeMarco, T. (1990, November). Making a difference in the schools. *IEEE Software*, 78–82.

Fowler, W. R. (1990, Fall). Cost justifying your training project. *Journal of Interactive Instruction Development*, 16–19.

Godkewitsch, M. (1987, May). The dollars and sense of corporate training. *Training*, 79–81.

Hammer, M., & Champy, J. (1993). *Reengineering the corporation*. New York: Harper Collins.

Kearsley, G. (1982). *Costs, benefits, and productivity in training systems*. Reading, MA: Addison-Wesley.

Lave, J., & Wenger, E. (1991). *Situated learning: Legitimate peripheral participation*. Cambridge, UK: Cambridge University Press.

Lesgold, A., Eggan, G., Katz, S., & Rao, G. (1992). Possibilities for assessment using computer-based apprenticeship environments. In J. W. Regian & V. J. Shute (Eds.), *Cognitive approaches to automated instruction* (pp. 49–80). Hillsdale, NJ: Lawrence Erlbaum Associates.

Lesgold, A., Lajoie, S., Bunzo, M., & Eggan, G. (1992). SHERLOCK: A coached practice environment for an electronics troubleshooting job. In J. Larkin & R. Chabay (Eds.), *Computer-assisted instruction and intelligent tutoring systems* (pp. 201–238). Hillsdale, NJ: Lawrence Erlbaum Associates.

Newell A., & Rosenbloom, P. S. (1981). Mechanisms of skill acquisition and the law of practice. In J. R. Anderson (Ed.), *Cognitive skills and their acquisition* (pp. 1–55). Hillsdale, NJ: Lawrence Erlbaum Associates.

Nielsen, J. (1993). *Usability engineering*. Cambridge, MA: AP Professional.

Senge, P. M. (1990). *The fifth discipline: The art & practice of the learning organization*. New York: Doubleday.

Papert, S. (1980). *Mindstorms: Children, computers, and powerful ideas*. New York: Basic Books.

Piller, C. (1994, October). Dreamnet, *MacWorld*, 96–105.

"Training and the Workplace: Smart Work," *The Economist* (22 August 1992), 21–22.

Underwood, B. J. (1954). Speed of learning and amount retained: A consideration of methodology. *Psychological Bulletin, 51*, 276–282.

VanLehn, K., Jones, R., & Chi, M. (1992). A model of the self-explanation effect. *Journal of the Learning Sciences, 2*, 1–59.

Woods, W. A. (1970). Transition network grammars for natural language analysis. *Communications of the ACM, 13*, 591–606.

Yeager, D. M. (1991). On-demand education: Transforming the way technical professionals work and learn. *IEEE 1991 Frontiers in Education Conference*, 294–295.

5

An Observational Study of Its Knowledge-Base Development by Nontechnical Subject-Matter Experts

Anne McClard
Media and Software Group
US WEST Advanced Technologies

The LEAP Authoring Cycle Study began during the spring of 1994. The project, an observational study of nontechnical subject-matter experts (SMEs) developing a knowledge base for the Learn, Explore, and Practice (LEAP) intelligent tutoring platform (Bloom, Bell, Meiskey, Sparks, Dooley, & Linton, 1995a), had several goals. Foremost, we wanted to know whether nontechnical subject-matter experts could understand the complexities of knowledge-base development and, with technical supervision, carry out the required tasks. We also hoped to gather technical requirements for the development of authoring tools to facilitate knowledge-base development for future courses. A third goal was to produce a formal authoring process that enabled the technical transfer of LEAP to an internal client organization at US WEST who would then be responsible for course development and maintainance of the system.

The following is a discussion of some of the findings from this study, focusing on issues that arose during the development process, especially those issues that likely need consideration during the development of a knowledge base for any intelligent tutoring system (ITS) to be deployed in industry. The first section, "The Knowledge-Base Problem," lays out the problems typically associated with the knowledge-base component of ITSs. The second section, "Some Findings From the LEAP Authoring Process Study," focuses on the process that the authoring and development team went through as well as pertinent issues. The third section, "Considerations

for ITS Course Development," expands on lessons learned about the authoring process as applied broadly to ITS knowledge base development and
includes some suggestions on what should be done differently in future ITS
authoring efforts. This section ends with a discussion of the feasibility of
using nontechnical SMEs for knowledge-base development. The final section,
"Epilog: Developing Tools Based on Experience," illustrates some of the
ways findings from this study have driven the initial design of authoring
tools for the LEAP platform.

THE KNOWLEDGE-BASE PROBLEM

ITSs have been around for the last 25 years (Wenger, 1987) and, yet, have
not made their way into widespread use in industry. There are a number of
reasons for this. Perhaps one of the most important reasons is the cost of
developing and maintaining such systems (Johnson, 1988). Additionally,
developing the knowledge-base component of any ITS is a complex task
requiring considerable involvement of subject-matter experts, instructional
designers, cognitive psychologists, and computer scientists (Fink, 1991;
Woolf, 1991).

The knowledge base is the component of an ITS that contains much of
the information that the tutoring system relies on to make expert decisions.
Other components of the system, such as the student and instruction modules, refer to the knowledge base to assess a student's relative performance
and to give appropriate advice and feedback. The knowledge base contains
all of the substantive information about lessons, topics, presentations, possible responses, and so forth (Woolf, 1991).

Knowledge bases, depending on the pedagogical goals and philosophy
behind the system, take a variety of forms. Some are more complex than
others. In the absence of authoring tools, all knowledge-base development
is arduous, requiring the expertise of technically skilled knowledge-acquisition experts aided by instructional designers, course developers, and other
subject-matter experts.

The knowledge-acquisition expert's role in ITS development is to acquire
and encode information obtained from instructional designers, course developers, and subject-matter experts. He or she must work closely with each
of these people to ensure that all information pertaining to the knowledge
domain as well as the structural relationships between its constituent parts
are understood, included, and mapped out. The integrity of an ITS rests on
its knowledge base.

Most ITSs include several modules: (a) the expert module, which contains
the domain knowledge (knowledge base); (b) the student module, which
diagnoses what the student knows based on his or her performance; (c) the

instructor module, which diagnoses deficiencies in student knowledge and makes adjustments accordingly; and (d) the instructional environment and human–computer interface, which channel tutorial communication (Burns & Capps, 1988). The functionalities of the student module, instructor module, intructional environment, and interface pivot around the knowledge contained in the expert module; they depend on its being an accurate representation of the structural relationships and pedagogical goals of the system (see Fig. 5.1a).

LEAP, as with other ITSs, includes an expert module (the knowledge base), a student module (student model), an instructor module (instruction manager), an instructional environment (the practice environment), and an interface (see Fig. 5.1b). The LEAP instructional environment consists of four parts: an application simulator, which simulates applications used by CCEs; a reflective follow-up module that provides "reflective opportunities outside of performance periods," and allows "users to reflect on the events of the customer contact they just practiced" (Bloom, Bell, Meiskey, Sparks, Dooley, & Linton, 1995, p. 38); and a contact selection module that provides a starting point for the simulated conversations that a student will practice. In addition to these modules, LEAP employs a dialogue manager, which "is responsible

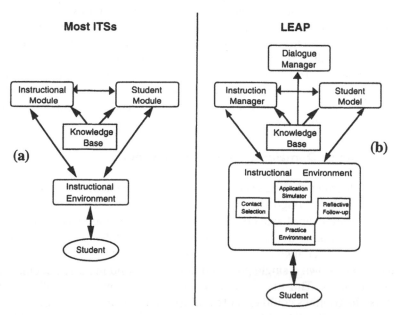

FIG. 5.1. (a) This diagram represents a generalization of ITSs. Some ITSs, like LEAP, include more modules, but nearly all ITSs have those represented here. (b) LEAP's instructional environment consists of the contact selection module, practice environment, application simulator, and the reflective follow-up module. LEAP also employs a dialogue manager.

for keeping track of the current location of the contact where the student is in the discourse grammar, and for making the appropriate transitions in the grammar based on the actions taken" (p. 38).

One confounding problem in ITS knowledge-base development has been the lack of adequate authoring tools for acquiring and representing knowledge for any single ITS. Ideally, the existence of such tools enables course developers, instructors, and subject-matter experts to develop and maintain knowledge bases without the support, or with minimal support, of computer scientists in a shorter period of time (Johnson, 1988).

SOME FINDINGS FROM THE LEAP AUTHORING PROCESS STUDY

Methods

Data collection for the LEAP Authoring Process Study spanned a period of approximately 17 weeks. The entire authoring process took 32 weeks. We formally collected data only during the first half of the development. Methods employed for collection of data included: participant observation, interviews, weekly questionnaires related to resource allocation, and videotapes of meetings. All interviews and relevant segments of videotape were transcribed. Nearly all of the members of the LEAP development team who had a direct impact on the authoring process and all members of the authoring team, including the product manager, were interviewed. Team members participated in interviews at least once every 2 weeks and submitted weekly time/task questionnaires.

The Knowledge Domain: Consultative Behavior

LEAP 2.0 (Bloom et al., 1995a) grew directly out of the field evaluation results from LEAP 1.0. Although the field trial was very positive, the developers concluded that several aspects of the first version needed to be changed to better serve training needs (Bloom, Linton, & Bell, 1995b). The most major and most critical change concerned dialogue flexibility. In version 1.0, trainees were "led" down a single path and had to respond in fairly specific ways at specific times. The hope was that by providing more flexibility in the contacts, the trainees could execute multiple actions or goals simultaneously and perform actions in different but acceptable orders, thus providing a more natural and realistic simulation of what CCEs do on the job. The developers thought that if they incorporated such flexibility into the system, LEAP would gain wider acceptance among new trainees. Such a system would also enhance acceptance among experienced CCEs engaged in ongoing training who had already established methods for handling customer contacts.

After the trial of LEAP 1.0, interest in LEAP grew considerably, and a new market unit expressed a desire to develop a LEAP course on "consultative behavior" to integrate with other training efforts at a large service center scheduled to open. "Consultative behavior" is the set of conversational skills that enable CCEs to confer with customers to meet their stated and unstated needs. Consultative behavior is now discussed in greater detail as it relates to how the SMEs structured the knowledge base.

SME Selection

In deciding what type of SMEs to request for the authoring of the LEAP course on consultative behavior, the developers needed to define what skills those team members should have. The skills they were looking for were: experience with LEAP 1.0, knowledge about developing course content for traditional training, knowledge about existing available training materials, knowledge of US WEST Communications culture, up-to-date knowledge of online systems, knowledge about consultative behavior, experience in both business and residential services, and political savvy. To some degree, luck brought together the individual experts that emerged as the authoring team.

The LEAP developers requested the help of a course developer, with whom they had worked on LEAP 1.0, from US WEST's training organization. They felt that she was an asset to the team because she was extremely familiar with CCE culture and training materials, and she was knowledgeable about course development. She also knew the political climate of the market units that were involved. The developers also asked for the help of a CCE, who had been involved with the development of LEAP 1.0, from a residential service office. She provided expertise with online systems and current policy and procedures for CCEs. Additionally, they enlisted the help of an expert in consultative behavior from the small business unit to consult with the team on content related to consultative behavior and to represent the interests of her organization. All three of the SMEs were long-time veterans of US WEST, and all three had worked as CCEs at one time or another.

Getting Started: Expectations

Initially, the members of the authoring team (AT) had little understanding of the scope or breadth of the project on which they were embarking. The consultative-behavior expert thought she was working on LEAP for a couple of weeks to consult on course content and then returning to her normal position. Both the course development and CCE experts anticipated a commitment of 6 to 8 weeks. It soon became apparent that the task laid out for them was much broader than they anticipated. Not only did they write the conversation scripts, but they were also responsible for breaking the scripts into structural components and putting them into coded form (the grammar, or knowledge base). For two of the SMEs, the duration of the project was a serious issue as they were commuting from distant cities.

At first, the team had difficulty defining their goals. This happened primarily because the client group had not defined their requirements clearly. Initially, it was difficult to get consensus among the various market units on what they meant by consultative behavior. Another constraining factor was that due to re-engineering, there was a great deal of flux in the involved organizations, in their objectives, and in precisely how everything was going to operate in the new service center (i.e., what applications were going to be used).

Despite these obstacles, the authoring team immediately began the work of authoring. As their work became more defined, so did their goals.

The Authoring Process

From the beginning, the authoring process was largely experimental. The development team did not know exactly what they could expect from the SMEs, nor did they have set ideas about how the SMEs should proceed. During the development of LEAP 1.0, a knowledge-acquisition expert developed the knowledge base. Unlike the consultative-behavior knowledge base for LEAP 2.0, the first knowledge base was intended to teach CCEs how to sell a particular product rather than to teach behavioral skills. The knowledge-acquisition expert spent hours with SMEs gathering information about how conversations might proceed, making sure that the information was an accurate representation of that domain. Rather than begin with detailed dialogue scripts, he began with abstract representations of conversations that CCEs might have, and from there, using a software tool developed for the purpose, he generated new conversations. The details of the scripts were completed with the help of several SMEs.

It was not clear whether the SMEs who were entrusted with authoring the consultative-behavior module of LEAP 2.0 could begin with the abstract representations that the knowledge-acquisition expert had, nor, given the domain of consultative behavior, was it clear that such an approach was appropriate. Left largely to their own devices, the SMEs came up with a plan on how to proceed. After ironing out a few details—who their target audience was, what the clients wanted, what holes existed in current training, and what was meant by consultative behavior—the authoring team began the painstaking task of writing dialogue scripts for the conversations that served as the foundation of the knowledge base.

Laying the Foundation.   The only requirements the authors had to begin with were that the content of the course focus on consultative behavior and that the content be appropriate for the business and residential market units. The consultative-behavior component was somewhat problematic. The client groups, although espousing the use of consultative behavior, had no clear definition of what it was, nor did many people understand how LEAP could be used to train it. After attending numerous meetings on the topic of consultative behavior, the authoring team still lacked concise information, so they drew on a variety of resources that had been used or were in use for initial training by the involved market units. From these resources as well as from information that the consultative behavior expert on the team provided, the authors developed a structured definition of consultative behavior. It was determined that consultative conversations contain several elements: establishing rapport with the customer, interviewing the customer to discover stated and unstated needs, presenting information to the customer, and closing the conversation in an appropriate manner. Furthermore, consultative conversations should begin on a human level, move to a business level, and end on a human level. Because CCEs, by company policy, must disclose very specific information to customers, the authors included this in their model of consultative behavior. Each of the aforementioned components of consultative behavior was broken down into its elements (see Table 5.1).

The authors learned that the target audience was CCEs engaged in ongoing training. For this reason and because all CCEs were required to attend in-class training covering consultative behavior, the authors decided that the course should contain only intermediate and advanced level practice materials. From their research of existing training, they came up with a list of topic areas for which CCEs needed more training. They used multivoting[1] to decide which of these presented the most critical need. They came up with the domain of problem solving. Using the technique of single-turn brainstorming, they then generated a list of seven problem-solving scenarios. For each of these, they wrote a customer–consultant dialogue incorporating the elements of consultative behavior.

The authoring team completed the initial work of authoring the conversation scripts early on, taking approximately 3 weeks to complete seven scripts and changing them for use in both business and residential services (see Fig. 5.2 for a sample script). While in the process of writing the scripts, the authors found themselves disagreeing about how complex the conver-

---

[1]Multivoting is a method used to help teams prioritize a lengthy list of topics. Typically, members of a decision-making group are each given a limited number of votes. Items on the list are ranked by number of votes received. Items receiving no votes are eliminated. The process is repeated until the list has been trimmed to an appropriate size.

TABLE 5.1
Components of Consultative Behavior

Topic	Definition
Rapport	Finding common ground with the customer
branding	• Saying the company name to anchor US WEST as a telecommunications leader with the customer
assure	• Informing the customer positively
affirm	• Validating the customer's decision
appreciate	• Recognizing the customer and expressing gratitude to him or her
acknowledge	• Letting the customer know you heard what he or she said
seek agreement	• Asking a question to determine if the customer is in agreement with you
basic telephone etiquette	• Common courtesies such as / Please/,Thank you/, You're welcome/ and/ May I put you on hold?/
offer additional assistance	• Checking to see if there is anything else you can help the customer with
customer service statement	• Advising the customer of US WEST's commitment to improve customer service
empathize	• Seeing things from the customers' point of view and being sensitive to his or her feelings
customer recovery	• Apologizing to the customer when US WEST has made an error or caused an inconvenience
Interview	Asking open, relevant questions to solve the customer's needs
explain procedure	• Explaining to the customer the process US WEST uses
clarify	• Seeking or giving more specific information
fact finding	• Seeking additional information by asking questions
verify	• Confirming or substantiating what the customer has said
reality	• Information gathered from the conversation, records, or market segmentation
feeling finding	• Seeking additional information by asking questions
Just Say When	• Seeking or giving the customer's desired due date
conclude	• Reaching a decision or agreement based on facts gathered
credit interview	• Asking questions needed to complete Credit Information
Present	Creating and sharing a solution to meet the customer's needs
confirm	• Verifying what the customer has said
explain procedure	• Explaining to the customer the process US WEST uses
recommend	• Presenting a solution for customer's consideration
customer recovery	• Apologizing to the customer when US WEST has made an error or caused an inconvenience

(Continued)

TABLE 5.1
(Continued)

Topic	Definition
just say when desired access	• Accommodating the customer's delivery time, also confirming access when necessary
full disclosure	• Government or company required statements that must be given to the customer
seek agreement	• Asking a question to determine if the customer is in agreement with you
cost	• Informing the customer of product/service prices as appropriate
transition	• Changing direction of the call to introduce a new topic
future pace	• Planting a seed for the customer to consider in the future
benefit	• Keeping the customer's perspective and emphasizing what's valuable to them
reality	• Information gathered from the conversation, records, or market segmentation
product satisfaction guarantee	• Statement or guarantees to ensure customer satisfaction
set expectation	• Letting the customer know what to expect
Close	Bringing the call to a conclusion
set expectations	• Letting the customer know what to expect
cost	• Informing your customer of produce/service prices as appropriate
branding	• Saying the company name to anchor US WEST as a telecommunications leader with the customer
Human Business	Human and Business aspects of conversations

Note. This structured definition of consultative behavior served as a foundation for the scripts, the topic structure, and the knowledge base of LEAP.

sations should be. Would CCEs really go the distance if they made it too hard for them? To achieve their pedagogical aims, they decided it was necessary to provide students with "good" and "better" examples of the same conversations. Good conversations represented a model of minimum achievement that they wanted CCEs to attain, whereas better conversations represented "the company line," the level of achievement expected by the com

Building the Knowledge Base: Conversations. As completion of the dialogue scripts neared, the development team decided to introduce the authoring team to the concept of the recursive transition network (RTN), the knowledge-base structure LEAP uses (referred to in LEAP as "the grammar"). The developers of the dialogue manager (DM), the part of LEAP responsible for

Script Examples Represented by Open Contact RTN

	Residential Conversation	Business Conversation	

Greeting-1	Phone: rings
	        <you-hear-a-tone>
		Phone: rings
		        <you-hear-a-tone>	

Greeting-2	CCE: USW Communications,
	      <USW>
	
	      This is Mark
	      <answer-with-name>
	
	      How may I help you?
	      <Offer-to-help>
		CCE: USW Communications,
		      <USW>
		
		      Small Business Group
		      <SBG>
		
		      This is Alma
		      <answer-with-name>
		
		      How may I help you?
		      <Offer-to-help>
			Greeting Sub-dialogue

Customer's-Opening-Statement	Customer: Hi, my name is Chan Lam
	      <give-name>
	
	      My phone number is 494-1968
	      <give-number>
	
	      I have a note to call you
	      <state-reality>
		Customer: Our phones aren't working.
		      <state-problem>	

Acknowledge-Customer's-Opening-Statement-1	CCE: (continued)	CCE: (continued)	**Acknowledge-Customer's-Opening-Statement-2**

FIG. 5.2. The beginnings of two scripts are illustrated here. The labeled circles on the side represent nodes or subdialogues, and the text underneath script items represents the action to be taken.

keeping track of where the student is in the grammar, met with the authors. They drew state diagrams that illustrated what an RTN for a sample conversation might look like (see Figs. 5.3 and 5.4).

The authors seemed to grasp the concept of the transition network well enough to map out individual conversations. In preparation for the work ahead, they began to look at the conversation scripts as they related to the structural definition of consultative behavior. They mapped these out on their scripts initially by using multicolored "Hi-Liter" pens, a method they later abandoned in favor of a table structure with text outlining.

Two of the authors continued working on the scripts, and the course developer on the team began to learn how to input object-node definitions (state definitions) into the computer. To facilitate matters, she created a macro so that she did not have to type in the static elements of the definition each time. Once she felt comfortable with the task, she taught the other two how to do it.

The team worked for a week in a distant city, away from the development team. Upon their return, after having completed the object-node definitions for all of the conversations, they discovered a significant misunderstanding. They could not treat each conversation individually. LEAP required the conversations to be contained in a single data structure (the RTN). Although it had been explained to them, they did not fully understand the concept of

Open Contact Recursive Transition Network

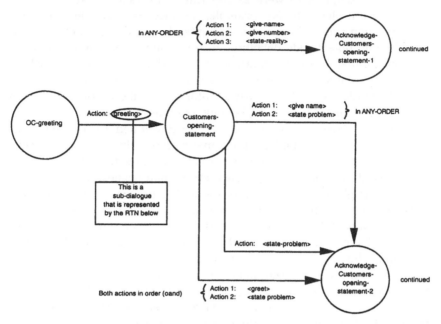

FIG. 5.3. The script in Fig. 5.2 is illustrated here as a state diagram. With this view, the transitions from one node to the next are more explicit. The subdialogue "greeting" is shown in Fig. 5.4.

Greeting Subdialogue from Open Contact

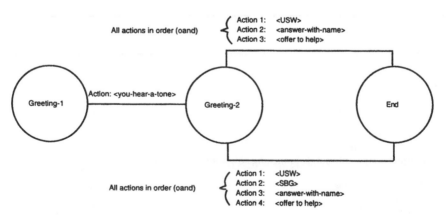

FIG. 5.4. When the subdialogue sequence is completed, the parth returns to the transition network above it in the subdialogue hierarchy. In this case, at the "End" node, one returns to the "Open Contact" network displayed above and continues from the "Customer's-opening-statement" node.

reuse. Even though there were 28 separate conversations, these shared many structural elements. For example, because each conversation contained an "open contact" sequence of events, this only needed to be represented by one object-node definition in the grammar as long as the node definitions pointed to the appropriate conversations. The rationale behind this system is that similar elements across conversations need to be linked so that the student model can abstract information about how the student is doing on similar tasks in multiple conversations.

Because the development team had never developed a knowledge base from the ground up (beginning with conversation scripts instead of beginning with the grammar), they were not really sure how the authors should do it and could provide little guidance other than suggesting approaches the authors might take. The authors began the process of "merging" the scripts into the grammar by trying to map the conversations out in state diagrams on a whiteboard. The diagrams soon became unwieldy, and they were forced to come up with some other solution. They developed a method using spreadsheets to map out conversations.

The task of merging the scripts into a single grammar was time consuming (see Table 5.2). Once done, they were able to enter the object nodes as they had done before. This time they faced other obstacles having to do with difficulty in understanding the concept of reuse.

The object-node definition contains several pieces of information that have to be defined: what the node name is; what the transitions are, which

TABLE 5.2
Time Spent on Authoring Tasks

Task	Time (person wks)
Front end analysis	1
Writing scripts	6
Revising scripts	6
Deciding on branching structure	2
Building the knowledge base:	
Identifying application simulations required	1
Authoring simulations	7
Merging scripts into single branching structure	7
defining states	2
Defining transitions	2
Debugging	7+
Recording audio	2

Note. This table represents time spent on tasks directly related to knowledge-base development. The authors also spend significant amounts of time in meetings, one the phone, traveling, and engaging in other company business during the course of the project.

include operator information and names of possible actions; what the next node is; and what conversation names are associated with the node. The part that was most difficult for the authors and the development team to get right were the action names. For example, in the "Open Contact" RTN (Fig. 5.3), one action name is "state-problem." "State-problem" represents a conversation action made by the CUSTOMER in multiple conversations. What the CUSTOMER actually says in those conversations is different but is represented by the same object-action definition. "State-problem" is manifested in four conversations as: "My phone is not working"; "I want to talk to a supervisor"; "My phone isn't working"; and "Our phones aren't working." This is the audio the student hears for these conversations. The same object action may be shared by multiple conversations as long as all of the aspects of its definition hold true. For this reason, the name one gives an action must be specific enough to identify what the action is but abstract enough to apply to more than one conversation. Getting to the right level of specificity was problematic.

Initially, the authors gave similar actions the same name. The problem was that in different contexts, the action definitions were slightly different. The names the authors had assigned were not specific enough. The dialogue manager could not distinguish between the actions that were labeled identically but defined differently. One of the developers explained the problem to the authors, and again, there seemed to be mutual understanding. The authors again worked in a distant city for a week, and when they returned, the developers discovered that the names they had assigned were too specific. The names the authors had assigned were specific to individual conversations, so they no longer represented similar actions in multiple conversations. Much of this problem arose because the authors received conflicting instructions from the two people responsible for developing the dialogue manager. Once the developers agreed on the level of specificity, they were able to explain it to the authors. On the third try, the authors succeeded at naming the actions appropriately.

After this, the work of completing the object-node definitions and object-action definitions went smoothly, although it was a tedious process. The authors had to type all of the information into the computer, insuring that parenthetical statements were enclosed. Trying to find open parentheses in the code was difficult to do visually, so one of the developers created a tool that enabled the authors to debug. The next major task for the authors was debugging. As it turned out, it was not a small task. Debugging took more time than all other tasks.

Building the Knowledge Base: Applications. At the point when the authors began the work of defining the object nodes and object actions, the CCE on the team began the work of authoring the application simulations.

She was the member of the team most familiar with the applications that CCEs were currently using and, so, was the logical choice for this job.

The first step in developing content related to applications was to identify which applications were needed and when. She determined this by going through the dialogue scripts and identifying which applications were needed as well as identifying what information needed to be input in specific fields of the individual applications. In addition to developing the content for the course, she worked closely with the member of the development team responsible for creating the application simulations. Her participation in this role was critical. For the application simulations to succeed, they needed to behave in exactly the same way as the real applications. She provided the developer with keystroke and feedback information that made this possible. In the process, she collaborated with him on the development of a tool to facilitate authoring the content for specific application simulators.

One major task in developing the content for the applications screens was data collection. The CCE authoring expert had to ensure that every detail was correct, including pricing of services and products for two states. The customer account information needed to be realistic, including credit information and history. To accomplish this, she collected actual customer records that corresponded to the dialogue scripts they developed. She modified this information so that it was a perfect fit with the details of the conversation. The application actions were subsequently tied in to the knowledge base.

Final Touches: Recording Audio. Recording the audio for the course took approximately 2 weeks and was one of the most straightforward aspects of knowledge-base development. The authors recruited individuals from the service center and from the development organization to play the roles of customers and CCEs. The volunteers read from scripts. The audios needed editing and needed to be named appropriately to allow the dialogue manager to find them.

Development Issues. The authoring team's introduction to LEAP consisted of a demonstration of LEAP 1.0. The team received no formal instruction or training on how to proceed with authoring, nor was this possible, given the situation. After all, the development team had little idea about how the process should proceed. Authoring team members complained at various times that they felt frustrated because they didn't know how how what they were doing fit together. As one member of the authoring team expressed it,

for as often as we have asked questions and tried to get a big picture of what's next and what does this look like and give us a time line, they don't seem to have it. So I, personally, am kind of frustrated with that. But that's part of what happens when you're doing development, I guess. I don't know. I know when we asked [X], before we left for Phoenix, we were told that [X] would be our contact ... So first thing we did Monday, called [X] in and tried to pick h— brain and get some stuff and some of it didn't quite match what we had understood and so later we talked with [Y] and, of course, [that] kind of reinforced what we were understanding and it didn't match what [X] had told us, and somewhere along the line I think the two kind of got together because then [X] came back later in the day and said I understand is somewhat in conflict with what [Y] told you. And so I think they're going through their own growing pains, themselves.

In retrospect, it would have been beneficial if the development team had spent more time in the beginning teaching the authors about RTNs. At that point, LEAP 1.0 existed and had a working grammar. Although that grammar differed in some significant ways from the grammar the authors needed to build, they would have benefited from the careful study of it and how it was assembled. The issues of reuse, with which they struggled in the process of developing their own grammar, may have been clarified had they understood from the beginning how an RTN works. It is also likely that their approach to building their grammar would have been slightly different.

The development team *was* "going through growing pains." The second version of LEAP incorporated new functionality that required a complete reconstruction of the dialogue manager, and it had undergone a complete interface transformation. The developers were working feverishly to complete work on the platform while the authors were developing the knowledge base. As a result, sometimes, developers were not available to help the authoring team when they needed them. Another problem created by the codevelopment of the knowledge base and the platform was that sometimes the developers were not sure how a particular functionality would be implemented or how it should be represented in the knowledge base, and, therefore, they did not always give the authoring team as clear information as they might have otherwise.

One benefit of the simultaneous development was that the authors were able to make significant contributions to the functionality of the LEAP 2.0 platform. Because they understood the ramifications of increased flexibility, they wanted to include a type of flexibility that the development team had slated for development in a later version. The authors wanted students to be able to perform certain actions throughout defined interims in the conversation. For example, CCEs vary greatly as to when they will enter certain information into applications, and the authors thought this should be supported. The development team consequently used the notion of a *fork* to accommodate this functionality.

The simultaneous development of the platform and the grammar was not ideal. Developers did not have the resources to dedicate one person to overseeing the authoring effort full-time or even half-time. Although one of the goals of the project was to see if nontechnical SMEs could develop a knowledge base, it was unreasonable to think they could do this without the constant support of technical staff. The development team was unable to provide enough support and guidance for them throughout the process, especially in light of the fact that the authors had been given minimal introductory information about the tasks they were required to accomplish.

Cultural differences between the development team and the authoring team were significant. The developers all came from research backgrounds, but the authors came from service backgrounds where they maintained more rigid schedules. Attitudes about time and manifestations of work ethics differed between the two groups. The authoring team worked very long days, often 12 hours, arriving at work at 7 a.m., and leaving at 7 p.m. Development team members, who also worked full days, long days when needed, approached time as something more flexible. Most of the development team arrived between 8 and 9 a.m., leaving between 5 and 6 p.m. When the job required extra work, they returned after dinner and worked late into the evening. The result of these differing approaches to time use was that the authors frequently worked unsupported and felt frustrated that the developers were not always around when they needed them.

Another manifestation of cultural differences initially was vocabulary. Development team members were used to communicating about LEAP in relatively technical terms, and authoring team members had little knowledge of this language. During the first couple of months of development, communication between the two groups was difficult, especially when it came to talking about the RTN and how it worked. Many of the mistakes made early on resulted from this inability to communicate efficiently. As the authoring team became more immersed in the project, they developed their own vocabulary for talking about the knowledge base among themselves. The terminology they used was still quite different from that of the development team. As a result, the two teams would leave a meeting thinking that they understood each other, only to find out later that they did not. The two teams eventually developed a common vocabulary.

Although they developed a common vocabulary, they never came to share an image of "the big picture." As development of authoring tools proceeded, this became more apparent. The authors' "big picture" focused on the details of conversations; they did not see what they were doing in the context of a large interconnected transition network. For the developers, on the other hand, the context was always abstracted to the grammar and where one was in the grammar. Again, had the developers been able to spend more time initially teaching the authors about RTNs, using relevant examples, the two teams would more likely have come closer to sharing this view.

For the most part, the authors were colocated during knowledge-base development. They shared a large work area and frequently asked each other questions. When application simulation development began, the developer responsible for the application simulations moved into the shared work area. This situation was ideal for both the developer and the CCE working on the applications. The few periods when the authoring team worked away from the development team were problematic. Although the two teams communicated by telephone on these occasions, communication was not as effective as it would have been had the two teams been colocated. Both times the authoring team worked away, they went down a wrong path, and the error was not discovered until they returned. Had the team been colocated with the development team during these two periods, it is likely that the mistakes would have been caught earlier than they were. An alternative solution to this problem was for the authoring team to have the ability to visually show what they were doing while talking to developers (for example, using a shared, computational workspace and telephone). At US WEST, as at many other large companies, teams often consist of personnel from disparate locations. Ideally, authoring teams should be situated together. Otherwise, the tools for developing knowledge bases should incorporate collaborative tools to enable individuals to author from their home bases.

One critical problem for the authoring team had to do with client specification of requirements, as mentioned previously. Not only were the specifications poorly stated at the outset of the project but they changed throughout the course of knowledge-base development. Each time the authors learned of a change in the curriculum plans for the new service center, they had to make changes to the scripts and, consequently, to the knowledge base itself. The changes they had to make to the knowledge base were cumbersome because they lacked adequate editing tools. Searching through thousands of lines of code was their only recourse in these situations. Had the requirements been set in the beginning, the authors would not have faced some of these difficulties. Nevertheless, as even in the best of all possible worlds things change, a need for tools that allow the knowledge base to be modified easily is indicated.

CONSIDERATIONS FOR ITS COURSE DEVELOPMENT

This section focuses on considerations relevant to the development of courses for any ITS, regardless of the platform or the structure of the knowledge base.

Clear Course Requirements Statement

As already mentioned, the lack of well-documented requirements for the "Consultative Behavior" course presented a major stumbling block for the authors. During the life of the project, they spent far too much time concerning themselves with a job that should have been done before they ever embarked on developing the knowledge base. Ideally, it was not their job to figure out for whom the course should be developed, to discover holes in current training, or to have to arrive at a definition of what consultative behavior is. Other problems, such as the shifting organizational structure, were unavoidable.

Before embarking on development of a knowledge base for a course, one should have a well-documented requirements statement at the outset of the process. This facilitates course development, taking the onus of researching and defining issues off the authoring team. It takes the guesswork out of the course development process, increasing the likelihood that the client will be satisfied with the product, and that the product will serve its intended instructional purpose. Another benefit of having a clear course-requirements statement from the beginning is that the overall cycle time of knowledge-base development will be reduced; authors can spend valuable start-up time on learning and planning activities, and development time will not be wasted on revision of already developed materials to incorporate newly discovered requirements.

Selecting Subject-Matter Experts

One does not always have perfect control over the selection of SMEs in industry. Much depends on who is available, and some organizations will not honor requests for specific people. In some ways, SME selection depends on the luck of the draw. The developers in this case were very lucky. Not only did they get the SMEs they requested but these SMEs also got along and worked together remarkably well. In selecting the SMEs for LEAP, the criteria were: some knowledge of LEAP, experience with development of course content for traditionally taught training, knowledge about the market units involved, knowledge of online systems, and knowledge of the domain.

The SMEs for development of any course should have firsthand knowledge of the topic areas covered in the knowledge base. Duplication of some skill sets is appropriate, but one should seek as much diversity on the team as possible. At least one person on the team should have prior experience with knowledge-base development, and the same or another person should have experience with instructional design. None of the LEAP authors had prior experience with knowledge-base development, and it is unlikely one will find such a person outside of a technical organization. Even with good authoring tools, it is likely that one member of a knowledge-base development team will need a comprehensive understanding of what such development entails. Other SMEs should be selected according to areas of expertise. All SMEs should minimally have keyboard skills on one type of computer. The LEAP authors were technically skilled when it came to the online systems used by CCEs but required some time coming up to speed on the systems used for development. If the SMEs do not have adequate computer skills, one must schedule more time for the development of these skills during the training period.

Training Period for SMEs

The lack of understanding about the overall task of developing a knowledge base plagued the authoring team throughout a good portion of the development process. Authors felt frustrated at not knowing how the individual tasks they were doing fit into "the big picture." It was also difficult for them not to know in relative terms how far they had come or how far they had to go. Throughout much of the process, the team struggled with the technical vocabulary largely because there were few familiar conceptual hooks to which they could attach their knowledge. They often used the same words as developers but understood them differently. They developed their own language about the knowledge base, a language the developers did not know. Many of the resulting communication problems may have been alleviated had the development team attempted to provide some training in the beginning.

In any knowledge-base development effort relying on nontechnical subject-matter experts, it is critical to provide a sufficient training interval before embarking on development, regardless of whether tools are available. The duration of the training period should vary according to the complexity of the ITS platform, its knowledge-base structure, and the nature of the course being developed. The goal of training should be to familiarize SMEs with the platform, to instill in them an unobscured understanding of how the system works, to teach them how to use extant authoring tools, and to teach them the vocabulary necessary to communicate effectively with technical support personnel. Furthermore, the training period should be used for team building, especially if the SMEs have never worked with one another previously. Initial training should take place in a single location even if the SMEs will be working in distributed locations during the development period.

Distributed Work

The fact that the LEAP authoring team was colocated with each other and with the development team throughout the majority of the project facilitated communication greatly. On those few occasions when the teams were not

together, productivity was less than optimal. When the authors were separated from one another, there was some duplication of effort, and when the
authors were separated from the development team, the lack of communication allowed the authors to head down wrong paths uncorrected. The
primary problem was not that they could not talk to each other but that it
was difficult to understand what the issues were without actually seeing the
results of the work that was being done.

If SMEs will be working separately from distant locations, they will require
adequate technical support and feedback systems. This may require added
infrastructure. For example, perhaps they will need two-way video conferencing and/or a shared computational workspace, allowing for simultaneous
modification of a single knowledge base. Colocation is likely to be necessary
during some phases of knowledge-base development, regardless of the tools
available for information sharing. During the authoring cycle study, approximately 10% of project time was spent traveling. Therefore, when team members
are working from distant locations, one must budget for travel accordingly.

Well-Defined Plan of Activities

At the beginning of the authoring process, the authors developed a plan for
creating the knowledge base that they recorded on a GANTT chart at the
request of the technical director. The problem was that the authors did not
really know what tasks they were going to do, what those tasks were going
to involve, or how much time each task was going to take. Furthermore,
because they did not know what was involved, they had no clear idea about
how the work might be divided to complete it in the most efficient way. The
project planning, perhaps, should have been left up to the development
team, especially because they were the only ones who knew from the outset
what components had to be authored for the knowledge base.

Any authoring team benefits from a well-defined plan of activities and
tasks to complete. Each member of the team should have a clear view of
his or her responsibilities on the team and how these responsibilities fit
into the overall plan.

Critical to Develop Good Authoring Tools

The observational study of the LEAP authoring process by nontechnical
subject-matter experts points to the absolute necessity of the development
of good authoring tools for knowledge-base development. Most course developers and subject-matter experts do not have the technical knowledge
to build a knowledge base from the ground up—the way the LEAP authoring
team did. Furthermore, most training organizations and companies do not
have the resources to support such endeavors. For LEAP, or any other ITS

platform, to be widely used in industry, the development and maintenance of the knowledge-base component needs to be more accessible to those people whose job it is to develop and maintain training materials for companies. Tools for authoring will greatly facilitate the process of knowledge-base development. Regardless of how good the tools are, they will not eliminate the need for authors to have a conceptual understanding of how a particular ITS works.

How Feasible Is Relying on SMEs for Knowledge-Base Development?

This experiment demonstrated that nontechnical SMEs are capable of knowledge-base development. Considering the adversity that faced the LEAP authoring team along the way, they did a remarkable job of overcoming obstacles and accomplishing the tasks they needed to. The knowledge base they developed is impressive from both technical and qualitative standpoints. The time it took them to develop the course, however, is unacceptable under any other circumstances. Were knowledge-base development to proceed so slowly in every case, it is not feasible to use SMEs for anything other than touching up the details. The development of good authoring tools will make knowledge-base development by SMEs more feasible. Nevertheless, even with the development of good authoring tools, it seems likely that at least one technically trained person needs to be on hand throughout the knowledge-base development process. As tools become more robust and easier to use and as expert users of such systems evolve, the need for technical support may lessen. The inclusion of nontechnical SMEs in the development of training materials is critical to training effectiveness.

EPILOG: DEVELOPING TOOLS BASED ON EXPERIENCE

The consultative-behavior knowledge base, although complete at the time of this writing, was still evolving. New training modules were being added as new training needs arose. Additionally, word about LEAP was spreading, and there had been several requests for courseware to meet other training needs in the company. To keep up with the demand for new training materials, the development team knew it was imperative to develop authoring tools that made the knowledge-base development process more cost effective by greatly reducing cycle time. To this end, even before the consultative-behavior knowledge base had been completed, the development team began to design authoring tools.

Some "stop-gap" tools were developed in direct response to needs expressed by the authoring team during development of the knowledge base. As mentioned earlier, one developer created a tool for authoring application simulations. Another developer made a tool to help debug the grammar. These tools served the immediate needs of authors but did not address many larger issues with which this authoring team wrestled. It seems likely that future authors will surely have to contend with many of the same issues. Armed with data from the observational study, the insights of the authoring and development teams, and a human-factors expert, we embarked on the collaborative design of the LEAP authoring tools.

One benefit of having documented nearly all phases of knowledge-base development with weekly interviews, time–task records, and videotaped meetings was that members of both the authoring team and the development team were forced into being more reflective about what they were doing: They had to keep track of how much time they spent on individual tasks, they had to think about what they did not understand and what frustrated them most in their work, and they were given the opportunity to reflect on their relationships with their coworkers and managers. Another benefit was that it was documented, analyzed, quantified, negotiated, and qualified (both teams agreed that the observations were an accurate representation of what happened). This reflectiveness on the process has guided the design process in significant ways, a few of which are enumerated next.

The study helped to identify and document areas of conceptual difficulty for the authors. Namely, authors had difficulty with the concept of reuse, evidenced by their difficulties in naming actions, nodes, and subdialogues appropriately. A related difficulty for authors was that they never came to "see" the knowledge base as an interconnected whole. Both of these conceptual problems presented serious impediments to the timely completion of the knowledge base. As a result, one of the foremost goals in designing the authoring tool was to make reuse and the knowledge-base structure more explicit to authors by using dynamic graphical displays of the grammar. As authors create or add materials, they see the the knowledge base building graphically. This representation makes the connections more obvious and cases of reuse more explicit.

Because debugging was such a major task for the authoring team, requiring more time than any other single task, the design team focused on developing tools that reduced errors. Two common errors were typing errors associated with multiple instances of references to the same objects in the grammar and unmatched or misplaced parentheses in the code. The tool is designed to minimize errors by reducing repetitive tasks, such as typing in the same information in multiple places and by automating tasks that inherently present risk of error. The user never needs to use programming syntax—this is done "behind the scenes." Where possible, required text fields are autopopulated with information that has been entered elsewhere. One of the most time-consuming aspects of debugging was finding the right instance in the grammar to edit. The tool design includes extensive filterable search tools that will greatly facilitate navigation through large grammars such as the consultative-behavior knowledge base.

One of the focuses of the authoring cycle study was task flow. We documented the task flow as implemented by these authors and then met to analyze the methods used. The bottom-up approach, beginning with the scripts and building the knowledge base from there, made some sense, especially for novices, because role-play creation would likely be a familiar domain for course developers. However, everyone agreed that the methods the authors had used (and the painful and time-consuming process of "merging" the scripts into the grammar) were not optimal. We wanted the tool to discourage this approach and to encourage authors who chose to use a bottom-up approach to build the scripts and the RTN simultaneously. Development of the first, simple conversation script requires little more than word processing. To add another conversation to the knowledge base, however, the user must deal with structural features of the grammar through direct manipulation of its graphical representation.

Finally, we wanted to move beyond the experiences of the present authoring team by including features and supporting task flows that an expert of the system might use. To this end, the tool is designed to support a top-down approach that allows the user to begin with the structure of the knowledge base, leaving the details to the end, just as the knowledge acquisition expert for LEAP 1.0 had.

The present authoring team's work strategies have changed considerably as they have gained experience. They currently use a combination of methods for updating, building on, and maintaining the consultative-behavior knowledge base. They have been active participants in authoring tool design efforts and have eagerly tested out ideas and prototypes. At present, the LEAP authoring tool is still under development. Until the authoring tools are tried and tested, it remains to be seen whether it is feasible to rely as heavily as we have on nontechnical subject-matter experts for knowledge-base development.

REFERENCES

Bloom, C. P., Bell, B., Meiskey, L., Sparks, R., Dooley, S., & Linton, F. (1995). Putting intelligent tutoring systems technology into practice: A study in technology extension and transfer. *Machine Mediated Learning, 5,* 13–41.

Bloom, C. P., Linton, F., & Bell, B. (1995). *Using evaluation in the design of an intelligent tutoring system.* Manuscript submitted for publication.

Burns, H. L., & Capps, C. G. (1988). Foundations of intelligent tutoring systems: An introduction. In M. C. Polson & J. J. Richardson (Eds.), *Foundations of intelligent tutoring systems* (pp. 1–20). Hillsdale, NJ: Lawrence Erlbaum Associates.

Fink, P. K. (1991). The role of domain knowledge in the design of an intelligent tutoring system. In H. Burns, J. W. Parlett, & C. L. Redfield (Eds.), *Intelligent tutoring systems: Evolutions in design* (pp. 195–224). Hillsdale, NJ: Lawrence Erlbaum Associates.

Johnson, W. B. (1988). Pragmatic considerations in research, development, and implementation of intelligent tutoring systems. In M. C. Polson & J. J. Richardson (Eds.), *Foundations of intelligent tutoring systems* (pp. 191–208). Hillsdale, NJ: Lawrence Erlbaum Associates.

Wenger, E. (1987). *Artificial intelligence and tutoring systems: Computational and cognitive approaches to the communication of knowledge*. Los Altos, CA: Kaufmann.

Woolf, B. (1991). Representing, acquiring, and reasoning about tutoring knowledge. In H. Burns, J. W. Parlett, & C. L. Redfield (Eds.), *Intelligent tutoring systems: Evolutions in design* (pp. 127–150). Hillsdale, NJ: Lawrence Erlbaum Associates.

6

Supporting Development of Online Task Guidance for Software System Users: Lessons From the WITS Project

Robert Farrell
Lawrence S. Lefkowitz
Bellcore

Interactive learning environments are moving beyond isolated applications in academic settings toward more generic systems with wider applicability in the corporate and commercial sectors. Our work has focused on the authoring of online task guidance to improve end-user performance on common job tasks. This chapter describes WITS,[1] a research and development effort to build a production-quality authoring system to support rapid development of intelligent tutoring systems (ITSs) for software applications. WITS Author™, the authoring system, was designed and tested in collaboration with members of our corporate training staff and allowed training specialists to build, without programming, an IT specific to a given software system. WITS Tutor™, the delivery system, allowed end users to access job-relevant training and receive personalized instruction as they performed tasks on the software systems that they use on the job. We have applied this technology to several Bellcore[2] telecommunications operations support systems[3] (OSSs) and have fielded one of these systems, a tutor for Common Channel Signaling (CCS) network message screening.

---

[1]WITS is the name of the project but is not an acronym.

[2]Bellcore provides consulting services and software systems to the telecommunications industry and is owned by the regional Bell Operating Companies.

[3]Operations Support Systems are the software systems used to support the daily operations of telecommunications service providers, including customer service, billing, provisioning, and planning.

We first explain the existing climate at Bellcore when we started the WITS project and the need for more effective training. We then outline the goals of the WITS project and the anticipated benefits. Next, we give a short history of the project and its major accomplishments. We also discuss the lessons we learned from building ITS authoring systems for use by a corporate training department and building ITSs for deployment in work centers. We conclude with some trends in the industry that have shaped our directions and a description of some ongoing projects at Bellcore.

BACKGROUND

The WITS project was a unique attempt to develop an authoring system for creating ITs for a large class of software systems. We set out to change the way that training was developed at Bellcore and the regional Bell Operating companies. We first review our company's need for alternatives to its methods of providing training on its software systems. We then look at the challenges involved in this project and some previous approaches to meet these challenges. Finally, we review the overall project goals and the anticipated benefits of deploying this kind of system.

Motivation

Bellcore has over 200 computer systems that are used to operate various telecommunications networks. The rush to become cost-effective by downsizing the workforce has resulted in an increasing market for OSSs among Bellcore's clients. The widening gap between basic education provided by the school systems and that required to operate these complex software applications has created an immediate need for high-quality training capable of being delivered when needed, on-site, in a cost-effective manner. Consolidation has forced employees to quickly learn several computer systems.

As indicated by the results of a survey of customers' satisfaction with Bellcore products (Tarr, 1990), the need for better training was not being met by current technology: Inadequate training for software products was consistently cited as the second or third most significant problem in the survey. The challenge for new training systems was clearly stated: "Training will have to be very efficient, increasingly more individualized, and on demand, embedded in the system that employees use to perform their work."

Challenges

Creating a training product for end users of Bellcore OSSs presented a number of challenges:

- *Scope*—Our original goal was to replace one or more days of classroom instruction at our Lisle training facility to reduce the client's cost of receiving this training. This meant that we had to support a range of instructional material.

- *Availability*—It was also important that the tutor be available on-site, on-demand, when it is needed, making it necessary to support shorter, more directed training episodes.

- *Variability*—The workplace is a much more diverse environment than the classroom. Students typically come to a classroom at a particular age (in the schools) or at a particular level of competence (e.g. when they are new at a job). In contrast, an on-site training system must effectively teach students of varying backgrounds.

- *Realism*—Training scenarios that do not reflect the actual work environment are not effective and undermine the credibility of the training organization. WITS training had to be built by subject matter experts and training specialists with field experience. It was crucial that the tutor accurately reflect the current version of the software being used in the field. The tutor had to feel and behave like the real system down to the error messages, prompts, and appearance.

- *Environment*—Work environments are busy and noisy, making it difficult to create instruction requiring lots of concentration.

- *Platform*—Computer hardware platforms, especially in the telecommunications industry, tend to be different than those in research labs or academia. We found it difficult to find customers with the proper hardware already in the field. Also, the work site may have strict computer usage requirements that prohibit running cycle-intensive training software during work hours.

Previous Approaches

At the beginning of the WITS project, Bellcore had some expertise in creating computer-based training (CBT), but mostly in facilitating standardization and sharing of CBT across the Regional Bell Operating companies. The popular delivery methods for training were stand-up courses offered at the Bellcore Training and Education Center in Lisle, Illinois, and "suitcase" courses where the instructor traveled to a local training center or customer worksite. Video courses, paper-based self instruction, and some IVDs (interactive videodiscs) were also offered.

The literature on CBT and the training department's experience at the time suggested that CBT was a mixed blessing. On the positive side, it was:

- *Efficient*—CBT can reduce the time to learn tasks by as much as 50% because the instruction is individualized. With the basic capability of branching, authors could add choices that allowed the student to cover only those topics that were new and those in which they were interested.
- *Effective*—CBT was considered effective for teaching simple procedural tasks such as keystroke sequences. It was not considered very effective at teaching complex problem solving and decision making.
- *Consistent*—All authors on a project or across the company could use standard formats and templates, making training more consistent.

On the negative side, CBT was:

- *Unpopular*—Feedback on CBT was quite negative. Most CBT was targeted at how to use "3270" terminal interfaces to IBM mainframe software. Multimedia-based instruction has made CBT more popular and engaging, but in most cases people still prefer to learn from a teacher.
- *Expensive*—Non-simulation-based CBT took somewhere between 100 and 200 hours per hour of instruction, and simulation-based CBT took even longer. Better authoring systems have addressed this problem to some degree, but simulation-based training remains difficult to develop.
- *Difficult to install*—Most training was installed with numerous floppy disks. Current CD-ROM and network deployment methods have alleviated this problem.
- *Unavailable*—Workers did not usually have personal computers at the worksite that could run the computer-based courses.
- *Out of date*—Simulations quickly became out of date as features of the system changed. Maintenance was expensive.

Because of these limitations, in 1988 only 3% of Bellcore's own OSS products had accompanying CBT. In addition, there was an unacceptable lag between product introduction and the availability of training on the system. Generally speaking, training took 6 months to develop. Work on training was started just after the product was released. By the time users received the training, a new release of the software was often available. The management of Bellcore's software organization and the top training officers in each of Bellcore's regional Bell operating company clients were ready to make a commitment to change.

Project Goals

We targeted the following kinds of training applications:

• *Remedial instruction and practice*—Originally, WITS was designed for classroom use with a teacher. It provided remedial instruction and practice for students that needed individual instruction beyond what could be provided by the teacher.

• *New product training*—A more cost effective method of using WITS was as a method of training users on new products. WITS was shipped along with the software product, and users could use it at the worksite.

• *Release training*—We had some desire to use WITS for training on additional releases of existing products. Because delivery was more immediate than classroom training, users could have some training available immediately when new features were released.

After some economic analysis, we decided that WITS tutors had the most benefit if we focused on new product training. This decision set our overall project goals and determined our overall direction. We decided that WITS-based new product training should be:

• *Developed with the product*—WITS authoring becomes part of OSS software development. Parts of the authoring process could start after requirements were completed, other parts after multiunit test, and the final portions during system test. This reduces any inconsistencies between the product and the training.

• *Delivered with the product*—As with online help and documentation, online WITS training is tested in the final stages of system testing and delivered with the product. The lag time between release and training availability is reduced to zero.

• *Accessible from the product*—Users can access the WITS tutor as needed, directly from the product or their computer's desktop, receive training on a particular issue, and return to their work.

• *Use real databases*—Users can take advantage of company-specific databases for training, instead of using Bellcore-supplied training databases. However, real data can be incomplete or corrupt, making it difficult to develop a robust training system that utilizes the data directly.

• *Real problems*—Problems needing tutor attention should be specified by the user or their supervisors, rather than predetermined by a training curriculum. Users should be able to describe the problems they are working on and receive immediate tutoring.

• *Run alongside the product*—UNIX workstations were quickly becoming the deployment machine for Bellcore OSSs. WITS ran alongside the OSS on the same platform. For operations systems running on mainframes with character terminals, it was too difficult to port WITS to a mainframe environment. However, when we began the WITS project, many of our clients

were acquiring PCs that could perform terminal emulation, making it theoretically feasible to run WITS on the PC and interact with a PC-based terminal
emulator.

Anticipated Benefits

When we started the WITS project in 1988, there were few studies available
that could quantify the results to be gained by deploying ITSs. Why were
Bellcore and the regional Bell operating companies willing to make a significant investment? By looking at the qualitative benefits provided by ITSs,
extrapolating from the quantitative gains shown in university and industry
studies, and considering the known costs of "business as usual," we predicted the following gains:

• *Less Time Away From the Job:* Time spent in training is time spent away
from the job. To obtain and maintain technical proficiency, employees spend
a significant amount of time in training each year. Decreasing the time required for training will have a significant financial impact. Studies have documented as much as a 50% decrease in training time for individualized ITS-
based training. Given, for example, a workforce of 10,000 employees, an
average employee cost of $50,000, and 10 days of training per employee per
year; a decrease in training time of even 25% results in annual savings of $5
million in "lost productivity" costs alone.

• *Decreased Travel Expenses:* Centralized training has many advantages
and is common in many organizations. Bellcore's clients must typically travel
either to the Bellcore Training and Education Center or to a regional center
for much of their training. Considering the same base of 10,000 employees,
and a rate of $1,000 per trip, savings of up to $10 million per year are possible.

• *Increased User Effectiveness:* Incorrect or suboptimal use of operations
systems can be extremely costly, both in terms of wasted effort to correct
the results and in terms of substandard performance based on these results.
Bellcore concluded that a 5% increase in user effectiveness on a high-lever-
age, widespread operations system could result in annual savings of $1.5
million. If WITS training produces the expected 5% to 25% gains in user
effectiveness and is applied to equally high leverage systems, gains of tens
of millions of dollars per regional company per year are feasible. However,
because we did no summative evaluation after we built our prototype, we
cannot accurately quantify the increases in effectiveness that are possible.

• *Decreased Authoring Time:* Simulation-based training can take as much
as 300 hours per hour of instruction to develop. We projected that WITS-
based courses take less than 200 hours per hour of instruction to develop.
We tracked actual course development time for a sample WITS course that
was developed as part of stand-up training on WITS Author. Our course

developer took only 80 hours per hour of instruction, a time comparable to that for stand-up training.[4]

Additional indirect gains could come from the improved representation that OSS users will have of the task knowledge that they are learning. We anticipated that this allowed them to apply their knowledge more easily to other jobs. Based on all of these advantages, Bellcore decided to invest in a generic platform for building ITSs.

HISTORY OF THE WITS PROJECT

The WITS project evolved from a single-person summer project into a full-scale development project. The production software consisted of the authoring system, WITS Author, and the delivery system, WITS Tutor. WITS Author allowed software system development groups and training organizations to construct ITSs for a broad class of mainframe-based software applications. WITS Tutor supported deployment of the tutoring systems created in WITS Author.

Prototype Systems

The purpose of the two prototype systems was to prove that ITSs could be built for OSSs and to generate interest in the technology. We did not receive funding to evaluate the prototypes nor to develop them into a product. However, our prototype efforts indicated that it was worth investing in an authoring system to more quickly produce these kinds of tutoring systems and to give the tools to those who know how best to create content.

The PICS/MOA Prototype. In 1988, an online training system for PICS (Plug-in Inventory and Control System) / MOA (Mechanized Order Acknowledgment) was developed as a proof-of-concept prototype. It provided an overview of the MOA application and hands-on practice in filling out forms This prototype served to demonstrate the utility of applying IT technology to OSS training. The work was funded as a summer project by the PICS group and work continued in the Computer Technology Transfer Division.

The LEIS Prototype. Beginning in 1989, an application-specific training prototype was developed for a portion of LEIS™, a planning and engineering OSS. We had the goal of providing a remedial instruction and practice facility for students taking courses at the Bellcore Training and Education Center.

---

[4]This number does not include front-end analysis of training needs.

We interviewed LEIS engineers and talked with end users of the system to determine requirements and the applicability of the IT approach developed in the PICS prototype.

The student problem solving could be divided into a decision-making phase, performed with pencil and paper, and a procedural computer-usage phase, performed using the LEIS computer system interface. In the first phase, users made layout and grouping decisions for telephone cable installation projects. In the second phase they input their solution into the LEIS system. At the time we started the project, no graphical front end had been constructed for LEIS, so we started by giving the user a view of the network that was identical to the paper and pencil view with which they were familiar and then provided a direct manipulation interface for inputting information. Because the students had to go back to their jobs and use the LEIS system user interface, we provided instruction on mapping the paper and pencil results to the computer system.

In the decision-making phase, there was no one correct answer, so we could not use an instructional strategy that guides the user toward a particular solution. Instead, we allowed them to specify a solution and submit it for evaluation. We identified a set of heuristics for evaluating solutions and for suggesting which steps were questionable and which steps incrementally improved the solution. Instead of simply pointing out the improvements to students, we modified our instructional strategy to ask students to justify their questionable step. We then gave them feedback on whether they chose the correct justifications.

The LEIS prototype could have provided a basis for a generic IT product: the students learned computer-usage skills by receiving immediate feedback on each of their actions based on a procedural expert model, as was done in the PICS/MOA prototype. In the LEIS prototype, however, the graphical user interface, the expert model, and much of the instructional strategy were domain specific. In addition, the LEIS computer system was modeled with a hand-crafted simulation. We felt that a simulation was too costly to develop and did provide adequate realism for the student. We decided to look for other applications and to center our efforts on making the computer skills component generic.

Deployed Application

Beginning in 1991, we started on a project to build a WITS-based training application for Bellcore's Gateway Screening Administration Tool (GSAT). This project involved many challenges as both the WITS software and the GSAT software were evolving during the training application development period. In addition, because GSAT ran on UNIX workstations and it had a graphical user interface, the WITS application interface code had to be generalized, and GSAT had to be modified to enable successful communication with the WITS tutor.

The training could again by divided into a decision-making phase and a computer operations phase. In the first phase, students generated a solution for screening unwanted messages from passing through a network gateway and were told if the solution was too permissive or too restrictive. Feedback was provided by generating messages that should be able to pass through the screen but could not or messages that should not get through the screen that could. These messages were simulated on a graphical display of the client's actual network. There was one correct screening solution for each problem. When the student found the screening solution, they were tutored through inputting the solution into GSAT. We also provided training on individual features of GSAT that might not be used when inputting a solution.

At any time during problem solving or going through lesson materials, the students could ask a question. Like Lang's P&Q interface (Lang, Graesser, Dumais, & Kilman, 1992), users selected from a set of menus to create a question. In our interface, the user first selected a question template and then selected objects to fill in the template from lists or by pointing to a diagram or a computer screen. We found that users did not perform question-asking per se with this interface, but they did find it useful as a way to browse the information in the knowledge base.

The lesson plan and other portions of the course were generic and developed concurrently with our authoring tools. Although the interface with the GSAT system had to be hand-coded, it was generic and we were able to support authoring of computer skills exercises and demonstrations using our new authoring tools. The decision-making portions of the GSAT course could not be adequately generalized, but they worked seamlessly with the other generic components. This training was completed and shipped with the GSAT product during 1993. Our work on this tutor is summarized in Lefkowitz, Farrell, and Yoo (1993).

Generic System and Authoring

Early in the project, we demonstrated our prototype system to Bell operating company upper management. This provided the visibility that was necessary to secure funding for our future work. In 1991, the Training and Education Forum, a planning and oversight group with representatives from each of the seven regional Bell operating companies and the Bellcore Training and Education Center, formed the Training Technology Task Force, which funded a broader initiative aimed at providing authoring capabilities and deployment on personal computers. Through 1992, we generalized the capabilities of the WITS tutor so that it could provide personalized instruction on any 3270-based application, and developed our initial suite of authoring tools.

In 1993, Bellcore formed a separate Learning Support organization aimed at providing online assistance for all Bellcore OSSs. The technology organization within Learning Support took a leadership role in creating a WITS product that could meet the deliverables to the regional Bell operating companies and form the basis for a commercial product. We developed an initial version of the WITS Author system that enabled trainers who were not programmers to develop WITS-based courses. We used this system, as it was being developed, to support development of materials for the GSAT course.

Starting in mid-1994, work began on a production version of the WITS software. The goal of this work was to increase the robustness of the WITS Author and Tutor and to provide the ability to deliver WITS-based training on PC platforms as well as on Unix™ systems. Toward this end, the WITS Tutor was largely rewritten with a new user interface developed in the Galaxy™ environment to obtain portability across Microsoft Windows™ and UNIX platforms. WITS Author also underwent several major changes based on feedback from usability sessions. At this point, the project grew to nine people, including project management, development, testing, training, and documentation. The WITS production release was completed by March 1994. We completed testing, ran a training class on how to use WITS Author, and started development of a WITS-based course on the Network Monitoring and Analysis (NMA) OSS.

Unfortunately, the delays encountered in porting to the PC platform, limitations in our approach to interfacing with the target software application, work remaining to support easy interfacing with GUI-based software applications, and limited multimedia capabilities as compared with CBT authoring systems compelled the Learning Support organization discontinue work on further commercialization of WITS. Bellcore corporate training is now working on task-oriented help, interactive documentation, and online training using standard CBT authoring systems. Our research has turned to embedded performance support systems, network-based delivery of interactive content, and sharing of expertise among a community of software users.

AUTHORING TOOLS

One of the major goals of the WITS project was to produce an authorable ITS for software applications. WITS had to be authorable to be competitive with traditional authoring approaches, to make course development consistent, less time consuming, and more methodical, and to be understandable and usable by our training department. We first discuss some of the issues facing designers of authoring systems for corporate users; we then describe the WITS Author system and how we addressed these issues.

Related Work

Authoring systems for interactive learning environments include knowledge-base authoring systems, hypermedia-based systems, and simulation-based systems. As with knowledge-acquisition systems for expert systems, knowledge-base authoring systems address the problems of knowledge elicitation and encoding. Murray and Woolf (1992) reviewed their development and evaluation of several knowledge-base editing and visual monitoring tools, including a tutorial strategy editor and a domain browser.

Hypermedia-based authoring systems automate the creation of educational hyperlinked multimedia. IDE (Russell, Moran, & Jordan, 1988) is an instructional design environment built on top of Xerox's NoteCards hypermedia system. IDE adds instructional capabilities to hypertext. For example, there are predefined instructional types such as "literature," "objective," and "principle." IDE had a simple domain model and a simple student model, and pedagogical rules for ordering topics and initiating instructional units. There are now many systems for creating educational hyperlinked multimedia for distribution over the World Wide Web, but they do not have instructional design principles built into them.

Simulation-based systems automate aspects of creating simulation-based ITs. Towne and Munro of USC Behavioral Technology Laboratories built a number of systems based on the STEAMER simulation environment (Hollan, Hutchins, & Weitzman, 1984). IMTS assisted its users in creating graphical simulations of complex machinery. RAPIDS extended this work by permitting an author to construct training scenarios using the graphical simulations built in IMTS (Towne & Munro, 1989). RAPIDS II's point-and-click editor allows authors to create "behavior" of new objects by combining the behavior of component parts.

There are some critical differences between these systems and WITS Author. WITS Author allows the user to construct an interface to a live system without programming. Most knowledge-base authoring systems are not generic to a wide class of software systems; they are specific to a given programming language or computer application. Hypermedia-based systems tend to be less structured than WITS Author and do not provide the same degree of interactivity. Simulation-based systems typically assume that the target system is completely inspectable, wheras a live software system is typically a black box.

Issues

When designing an authoring system, one needs to think not only about the information needed by the system being authored but also about what user population is providing that information, what processes they want to use to input that information, how they want to view it, and whether standards or reusable information already exist.

Target Users. One of the principles of user-centered design is to know your users, but this is not an easy task for the designer of an ITS authoring system. Is the primary user the domain expert who provides the necessary expert knowledge or the trainer who provides the requisite pedagogical knowledge? If one provides separate tools to these two groups, how does the knowledge become integrated into a coherent whole? In developing WITS Author, we found that the training organization was more interested in authoring than were our domain experts (software developers). The trainers wanted better ways to deliver their training and were willing to learn the system. Thus, we ended up targeting the system at the trainers and supporting them in inputting the expertise of the domain experts. The lesson in this is that the user population for ITS authoring systems is not always clear: you must interact with the target users and determine if they are willing and able to use the system.

The Authoring Process. The authoring process for an ITS usually mirrors the parts of the tutoring system: the simulation and/or interface, the student model, the teaching model, and the expert or domain model. Which of these should be authored first and which should shape the authoring process? Initially, we hoped to derive most of our authoring from the expert model. We found, however, that trainers were uncomfortable with the process of representing this knowledge. At first, we thought that this was because they were not very technical; most trainers did not understand the software domain as well as our experts. However, after some usability testing, we learned that their discomfort was actually with the methodology for knowledge entry that we had built into the authoring system. Trainers were uncomfortable with inputting domain knowledge when they could not see where and why it was going to be used. They needed a view of how this knowledge was going to eventually fit into the course.

Control Over Presentation. Any interactive system that makes presentation decisions takes some control away from the author. Still, allowing the tutor system to make presentation decisions was a paradigm shift for which the training department was not completely prepared.

We tried to involve the training staff directly in the design of the authoring system by doing some participatory design. An experienced course developer from Bellcore Training and Education Center sat with us in a usability laboratory for 2 days, going through authoring scenarios and using a paper "interface toolkit" to design the authoring system (Muller, 1991). Based on these sessions, it became clear that there was a difference between the trainer's conception of what needed to be authored and our conception. These differences came from both our overconfidence in the power of our model and from her reliance on traditional instructional design methodology for understanding what we were trying to do.

Originally, we viewed the tutoring system as an extension of the course developers' expertise. Instead of designing courses ahead of time, the course developers would be designing the courses to suit the students' immediate needs. However, this course developer wanted to have more control over the development of the course at "authoring time"; she did not want to relegate the important decisions to run-time as was recommended by the WITS team. She viewed our system as a standard CBT system where she should be able to specify the whole presentation in order. When we presented the tutoring paradigm, she considered it a competing paradigm that did not take advantage of the large amount of expertise that she and other course developers had built up in designing traditional courses.

One remedy to this problem was to construct a simple tutoring system that had little control structure, then use the expertise of the domain-expert trainers rather than the course developers to decide which topic to present next. If more aspects of the tutoring control structure were authorable and accessible to the trainers, as was done by Murray and Woolf (1992a), they might have taken some sense of ownership for those strategies. However, it is unlikely that the tutoring strategies and content planning done by current ITSs are robust enough for these trainers to trust.

After some prototyping, it became clear that our best strategy was to let students themselves have control over topic selection. Trainers were relegated to authoring the content and structure of the lessons without specifying the actual order in which topics were covered. The lesson plan served as an outline of the complete course. As in hypertext systems, the students were in charge of sequencing. The students could also interrupt any ongoing activity and return to the lesson plan. Interruptions were particularly important to handle during time-varying media such as animations, audio, or video (Muller, Farrell, Cebulka, & Smith, 1992).

Further work needs to be done to create an authoring paradigm that is acceptable from both a course-development point of view and a tutoring point of view. Perhaps a compromise is possible where the course developer's knowledge is used to suggest what topics to explore without exercising too much control over what the student actually does.

Knowledge-Base Standards. Hypermedia systems have become popular partly because of the degree of data standardization. In contrast, there has been little standardization in the representations used by ITs and interactive learning environments. One of the challenges in creating the WITS Author system was creating each of the knowledge representation formats that were needed for compatibility between the authoring system and the delivery system. Part of our problem was simply representing multimedia in a way that we could deliver a dynamic presentation that appeared to be hand edited. Now standards for tagged multimedia documents and better document manipulation languages make it easier to do this kind of presentation, but the problem also extends to how to represent procedural information in the domain model, lesson plans, dialogue structure, and more. The problem was exacerbated by the fact that we needed links from the domain knowledge to the multimedia information, and vice versa. We ended up with very limited connections between the domain knowledge and the instructional materials. In designing an ITS, a representation scheme using open standards that allows for integration of knowledge and other sources of training information should be adopted or created.

Viewing the Knowledge Bases. The classic problem for knowledge-acquisition systems in general and authoring systems for ITS in particular is that users do not always think in the terms used by the knowledge representation. One way to avoid this problem is to have users participate in the design of the knowledge base, but this approach is probably unlikely to succeed if there is a diverse audience of authoring system users. The authors should never need to see the knowledge base directly: they should always have a logical, use-specific view of that knowledge. Providing users with a WYSIWYG (What-You-See-Is-What-You-Get) environment for inputting knowledge reduces the barrier between the author and the system. In WITS Author, we hid the underlying knowledge bases from the users and gave them views that combined pedagogical knowledge and domain knowledge.

Reusable Knowledge. In the WITS project, we were confident that we had designed a knowledge representation for describing software systems that could meet our needs. We wanted to have the trainer input this information once, then use it in multiple instructional activities. For example, we provided a point-and-click way of entering information about the purpose and definition of OSS user interface objects. Unfortunately, the context in which this information is created made it useful for answering the student's questions, but not appropriate for providing assistance during exercises or for creating a multiple-choice test, as originally desired. We could have circumvented some of this problem with better natural language capabilities, but the problem was essentially pragmatic: in recognition tests, you don't want to give away the pairing between the question and answer. Trainers, not knowing that this information was to be used for a test, would enter "giveaways" (e.g., Q: "What does the Format key do?" A: "It formats the text on the screen"). The lesson for ITS authoring system design is to make the uses to which the knowledge will be put clear at knowledge entry time. For example, one might show samples of different ways the knowledge will be used.

Knowledge Entry. How does the user specify the knowledge in a form that is general enough to be used to drive the ITS? We had this problem when trying to get users to input general procedures for using the OSS. One approach that we tried was to have the users demonstrate how to perform the procedure using an example and then use our authoring tool to view the resulting steps and "interactively generalize" those steps into a general-purpose procedure. Another approach was to use very general prompts. Ultimately, we need ways of combining these approaches so that users are comfortable with the knowledge-entry dialogue. The lesson is to provide a consistent knowledge input strategy across the authoring system that forces the knowledge to be specified in a general manner.

The WITS Author Course Development Environment

The WITS Author system allowed trainers to build and test their course without leaving the authoring system (see Fig. 6.1). When they were ready to deploy a course, authors could package their course for deployment by generating an application-specific version of WITS Tutor that incorporated their course knowledge-base and accessed their lesson materials. The re-

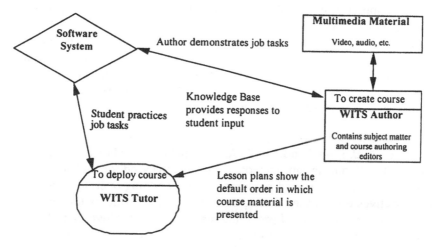

FIG. 6.1. WITS system architecture.

sulting ITS was installed on the software application's Help menu so that when the software was shipped to the end user, the training was immediately available.

Roadmap. To provide the author with a consistent and understandable authoring flow, we provided an initial "roadmap" that allowed access to each of the separate authoring tools (see Fig. 6.2). The roadmap provided access to two classes of tools: subject matter authoring tools and course authoring tools.

The subject matter authoring tools provided common information that was then viewed, annotated, and reorganized in the course authoring tools. Authors liked the roadmap because it suggested a top-to-bottom and left-to-right flow without forcing that flow on them. However, they thought that the roadmap window was too big and got in the way when they were using more than one authoring tool.

Subject Matter Authoring Tools. The subject matter authoring tools included the descriptions editor and the procedure editor. The descriptions editor allowed the author to connect to a mainframe-based software application and "capture" its screens one by one. Each screen was turned into a graphical display in the editor, enabling fields to be selected. One could also identify static parts of the screen to enable WITS Tutor to recognize the screen and do error correction. The author could also build up a knowledge base of information about the screen.

The procedure editor allowed the author to create task models to be used in other parts of the system. As the author performed a task on the target computer system, WITS Author displayed the steps in the procedure editor (see Fig. 6.3).

After the recording was made, it could be viewed, edited, and generalized. Finally, users could utilize the procedure as a demonstration of how to perform a common set of operations, a guided practice that took the student through the procedure step by step, or an exercise where the student had to perform the task without instructions.

Course Authoring Tools. WITS Author included facilities for supporting good instructional design, including an objectives editor and lesson plan editor. The authors created a course by first defining overall objectives in the objectives editor, then linking exercises, as assessments for that objective, and topics, as ways of providing instruction to achieve that objective. Finally, the authors used the lesson plan editor to hierarchically organize topics and their corresponding instructional activities (see Fig. 6.4).

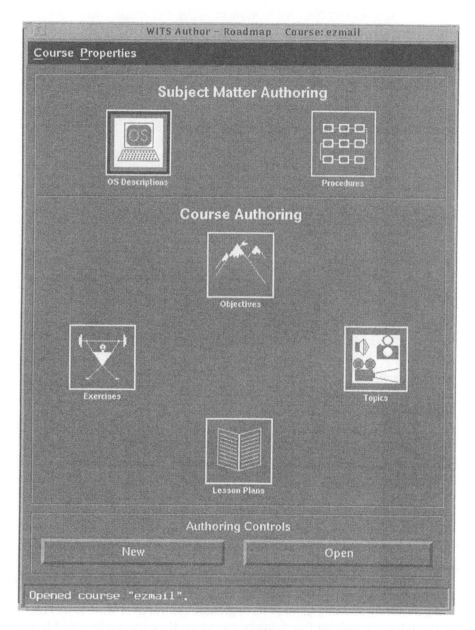

FIG. 6.2. The roadmap to WITS authoring tools.

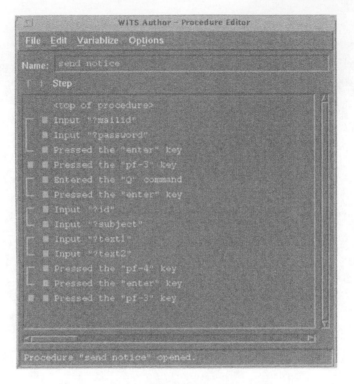

FIG. 6.3. The procedure editor.

Summary

One major lesson from our work on the WITS system is that our student population, author population, and deployment environment were quite different from those of most ITSs in the literature. Many of these differences didn't become apparent until we spent considerable effort building an initial version of our product. We suggest that anyone creating an authoring system for an ITS create a prototype that will give authoring users a concrete feel for what it is like to create a course, and then listen carefully to what they have to say.

We also learned that users must feel they have control over how the knowledge that they are inputting will be used by the delivery system. This can be done by allowing them to have greater control over what is included in the tutor, simplifying the delivery model so they can understand it, or giving them samples that illustrate how the resulting interaction might appear to the student.

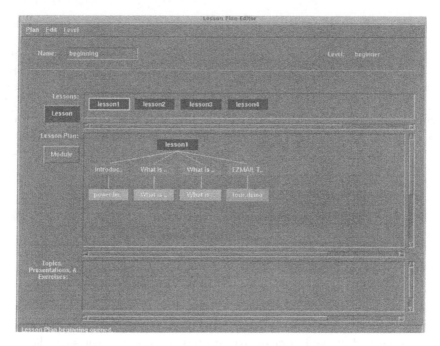

FIG. 6.4. The lesson plan editor.

THE INTERACTIVE LEARNING ENVIRONMENT

The WITS Tutor was a generic course delivery system combined with a practice environment that worked with the live computer system. We first discuss related work on generic ITSs, then we discussed some of the issues that we faced when designing WITS Tutor and deploying it at regional Bell Operating company work sites.

Related Work

Generic ITSs have been created for teaching troubleshooting and maintenance of complex electronic devices (Massey, Bruin, & Roberts, 1988) and programming (Anderson, Boyle, Corbett, & Lewis, 1990). However, there are few generic tutors for software systems. There are numerous generic tutors for text editors and word-processing systems, including the generic Tutor N Trainer (Egan, Nakatani, Shaw, & Hawley, 1987) that used synthesized speech to teach about the UNIX screen-based text editor, vi. However, the only ITS we know of that generically addresses form-based computer systems is LEAP (see Bloom's chapter), which used WITS technology to interact with the operations system.

There is ongoing work in generic shells in both corporations and academia. Early work included the GUIDON system (Clancey, 1987), which provided intelligent, computer-aided instruction using an EMYCIN-based expert system as its source of domain expertise, and byte-size tutor, an object-oriented shell for creating tutors with a direction–manipulation interface, which grew out of Bonar's work on teaching programming (Bonar, Cunningham, & Schultz, 1986). SIIP, a Self-Improving Instruction Planner, is a generic shell built on top of Hayes-Roth's blackboard architecture (MacMillan, Emme, & Berkowitz, 1988). Merrill built a number of course-development expert systems based on his instructional design methodology (Merrill, 1989). None of these systems have explicitly addressed the problems of a generic IT shell for software applications.

The WITS Tutor Delivery Environment

We focused on the following technical issues when designing the tutor:

- *Control*—The tutor had to effectively combine different levels of interactivity and student control. We wanted to support browsing, didactic presentation, user-initiated textual queries, and selectable graphics.
- *Knowledge structuring*—We wanted to support structuring of information by lesson plans, question/answer pairs, and other methods.
- *Integrated learning of concepts and procedures*—We wanted the tutor not just to present concepts and teach computer operations, but to integrate these two forms of knowledge in exercises that required problem-solving skills and computer skills.
- *Reuse of a common knowledge base*—We wanted all information available in the tutor—parts of lessons, answers to questions, feedback during problem solving—to come from the same base of information.

Like most ITSs (Polson & Richardson, 1988), we can divide the system into the following four modules: Interface module, Expert module, Student module, and Tutor module.

The Interface Module. The interface module is the IT's connection to the student (Polson & Richardson, 1988). Typically, there is one interface module that mediates all user inputs and system outputs. WITS Tutor used a generic multistream event architecture that handled input events coming from the tutor's graphical user interface and (potentially) multiple software applications. This section describes WITS Tutor's graphical user interface and its interface to the live software application.

FIG. 6.5. The WITS tutor main window.

User Interface. The graphical user interface (see Fig. 6.5) allowed the student to control the tutor by asking to abort the current activity, go on to the next activity, ask a question, or control various windows. It also displayed instructions, feedback, and other textual prompts and responses by the system.

As with most products with a graphical user interface, we went through several iterations with usability testing at each stage of implementation. We started with a "tiled" interface with areas on the screen for graphical explanations, textual self-study materials, and the software application simulation. Next, we changed an interface with overlapping windows where the self-study text window was accompanied by a lesson plan window. Individual media windows and the software application window were popped up and down as needed. Usability testing showed that users did not want to "manage" windows (pop the right window to the top, move windows around) and that we had too many controls available to the user. Finally, we implemented a Galaxy interface that consisted of separate windows but that had default locations tiled around the software system window. The main window was very small and fit just above the computer system terminal emulator window. We found that it was best if the tutor interface was as small as possible so that it could be displayed simultaneously with the computer system.

Application Interface. WITS maintained a live connection to the running software application. There were three key requirements driving our design of the WITS application software interface:

- No modification to application software source code.
- Support for menu-driven, command-driven, and form-oriented interfaces.
- Support for software applications on multiple platforms and user interface enviroments.

Our biggest requirement was that the target software application did not have to be modified in any way. Simple changes to large mainframe-based software can turn into thousands or millions of dollars in testing, documentation, training, and deployment costs.

Another requirement was that our tutor had to be able to work directly with live Bellcore OSSs. These software applications update databases, control network equipment, do forecasting, and so on. Based on a sample of several operations systems, we targeted menu-driven interfaces, command-driven interfaces, and form-filling as typical operations.

Finally, our application interface had to handle software applications running on different platforms (mainframe and workstation) and using different user-interface toolkits. Our customers had both large IBM 3270-based mainframe applications and small X-window workstation-based applications. We developed a generic user-interface action language that shielded the other modules of the tutoring system from the specifics of the toolkit in which the software system was implemented and the platform on which it ran. We supported the IBM 3270-based systems by modifying an in-house terminal emulator to output the user interface action language and then requiring the students to use that terminal emulator to access the mainframe. For the workstation-based systems, we modified the user interface toolkit used to program the OSS to output the user interface action language. We were not able to develop the toolkit modifications in time, so our deployed GSAT application was modified to output the user interface action language directly. Standardizing on the user interface action language early in the development of the generic tutoring system made the development of the other modules much easier. It is important to identify the nature of the application interface early and make alliances with groups developing the toolkits or environments in which the application will operate. The best approach is to build this sort of support directly into the operating system so that any advisory system, help system, or tutor could track application usage at a high level.

The Expert Module.   The expert module of an ITS represents the knowledge of a domain expert (Polson & Richardson, 1988). This knowledge could be procedural knowledge representing how to do tasks, or declarative knowledge representing domain facts and relationships.

A task model is a procedural domain model that represents the plan-based behavior of the expert or ideal student. A task model should be able to:

- Generate expert or ideal behavior.
- Serve as a model for explaining expert or ideal behavior to a student.
- Serve as a starting point for modeling student behavior.

In WITS Tutor, our task models represented the step-by-step procedures for carrying out common goals in the domain of a particular software application. For example, typical procedures for a mail application might include replying to a message, sending a message, and deleting a message. Each procedure consisted of a sequence of steps. Making the procedures both powerful and authorable was a challenging task. We developed a procedure description language that included looping and conditional execution that modeled some aspects of expert computer system usage. We found that it was too abstract for our trainers to author, so we ended up greatly simplifying our language. Providing adequate task models that can be understood by nonexperts is a crucial problem for the knowledge-base approach.

The Tutor Module. The tutor module supported both the instructional interaction during exercises and during self study. During exercises, WITS Tutor gave immediate feedback as the students attempted to perform a set of steps on the computer system.

The WITS tutoring method is based on "model tracing," a technique first used in the LISP Tutor (Anderson et al., 1990; Farrell, Anderson, & Reiser, 1985). The basic idea behind model tracing is that an ideal model of student behavior can be used to generate a set of paths for the student to follow. This ideal model is usually in the form of production (IF–THEN) rules, but any task model that generates a task flow can be used. At each step, if the student's input matches a path specified by the model, the tutor follows the student down the path. If the student's input does not match a known path, the tutor provides immediate feedback. Model tracing has been used successfully in a wide variety of ITs, including GRACE, an IT for COBOL, developed at NYNEX Science and Technology (Atwood, Burns, Gray, Morch, Radlinsky, & Turner, 1991; Gray & Atwood, 1992), and various ITs for high school mathematics developed by John Anderson and his group at Carnegie Mellon. We found that model tracing was an effective technique for teaching people how to operate complex software systems.

Self-study materials in WITS were organized into "activities" that could be run to create presentations to the student. Activities were organized into topics, which were reusable across lesson plans. Lesson plans were hierarchical organizations of lessons, modules, and topics. Activities included various types of multimedia presentations, demonstrations of software application behavior, guided practice in using the system, and exercises. There were also special optional activities for introducing and summarizing lessons. By following the structure of traditional instructional materials, our students were able to understand and follow the tutor more easily.

At the time we began to develop WITS Tutor into a commercial product, we could not find a multimedia development environment that allowed our authors to create multiple media that could be easily integrated with WITS Tutor lessons. We ended up supporting separate tools to display text, graphics, postscript, audio, and video. Where possible, we picked a cross-platform, popular media format and supported at least one editing program and one viewing program. We allowed our authoring customers to customize their multimedia setup by picking different editors or viewers.

Unfortunately, our multimedia approach resulted in a display that was:

- Difficult to follow because of the different windows popping up and down.
- Hard to manage because each tool had its own idiosyncratic control.
- Unsynchronized because separate media could only be displayed sequentially or started in parallel.

Because of these difficulties, almost all the self-study materials in the GSAT product were put into Framemaker™ and displayed in Ghostscript, a postscript previewer. The result was that despite useful exercises and interesting animation, students were quickly bored with page after page of noninteractive text and graphics, mostly taken from existing course materials.

Attempting to integrate multiple, separate viewing tools is difficult and not very effective. We recommend that ITS products either use a single multimedia environment or use a programming environment that allows embedded multimedia objects. Popular Web browsers now provide this kind of media-rich environment at low cost.

The Student Module. The student module of an ITS usually has the job of diagnosing errors or misconceptions and forming a model of the student's knowledge, skills, or goals that can be used by the tutor module to better target instruction. Various approaches to student modeling have been presented in the ITS literature (Sleeman & Brown, 1982) including the "overlay" model (Goldstein, 1982; Nwana, 1991) in which the student's knowledge is assumed to be a subset of the expert's knowledge, and "bug libraries" which enable a tutor to recognize common, stable classes of student errors (Sleeman, 1982).

In WITS Tutor, we used the "bug-library" approach. We classified *bugs* as specific, common differences between the task model and the student's input. Bugs could be specific to a procedure or applicable across procedures. These bugs were patterns of errors that we observed in computer system usage and did not require a root cause analysis.

We found that authors did not always have time to analyze user errors and come up with application-specific bugs, so they relied on our generic bugs. Despite their generality, our generic bugs were quite effective because they displayed context-specific information at a time when the student was having trouble and, thus, was ready to learn. It is important to anchor advice in the context of the error or question so that students can relate it to what they were doing. We did this by providing variables that could match against the student's input, the expected input, and aspects of the problem situation.

WITS Tutor used an attribute-based user model to represent different levels of student experience and various background information. The tutor module used this information to select an appropriate lesson plan or select appropriate feedback during exercises.

Several researchers have found that teachers and tutors rarely infer detailed user models from output traces. Instead, users tend to volunteer what it is they are trying to do or what it is they don't understand and tutors ask follow-up questions to clarify. We decided to let users access and update aspects of the user model at any point during tutoring.

Open Issues

A suitable interface to the live application is necessary for the tutoring system to adequately understand what the user is doing, but it is also important for initializing problems, noticing and recovering from errors, and tracking task completion. The application interface provides the basic vocabulary for both the student model and the task model. We first explain why live applications are to be differentiated from simulations in terms of the interface to the rest of the tutoring system. Then we cover several open issues: how to use the history of the user's action on the user interface, get the right level of granularity, and capture the human expertise at recognizing states of the software application.

Simulations, Black Boxes, and Live Applications.

There are two major approaches to interfacing a tutor, advisor, or other adjunct to a live system or a simulation: the glass-box method, in which the adjunct has complete access to the live system or simulation; and the black-box method where the adjunct can only examine inputs and outputs from the live system or simulation. Our purpose in using these terms is to draw an analogy with black-box and glass-box experts (Polson & Richardson, 1988). A black-box expert can simply return answers; it cannot explain its answers or provide access to its reasoning trace. A glass-box expert is open to inspection and can provide explanations of its reasoning.

There are two classes of glass-box application interfaces: inspectable simulations and completely open software systems. Inspectable simulations are directly accessible by the rest of the system. For example, the expert module can directly utilize the state of the simulation to determine the expert response. STEAMER (Hollan et al., 1984) was an example of an inspectable simulation. Completely open software systems are those that allow inspection of software state during system execution.

Inspectable, glass-box simulations are used most often in the ITSs literature. As a simulation, they have the advantage that they are resettable, and
their actions are reversible. This means that an expert module interfacing
with a simulation can utilize the simulation to search in a problem space:
the tutor can execute its expert model, examine the resulting state, and
undo it if necessary. Also, students that run into difficult states can be
backed out of these states by reversing or resetting the simulation. As a
glass-box interface, it is easy to get accurate data about the state of a
simulation. Getting data from a live system may be slow, error prone, or
impossible due to bad connections. Our early attempts utilized glass-box
simulations of IBM 3270-based user interfaces because simulations of small
parts of a systems are easy to build. However, even with such usable tools as
those provided by IMTS (Towne & Munro, 1988), authoring glass-box simulations can be very time consuming and, hence, prohibitively expensive.

ITSs can have a black-box application interface to either a live application
or to a simulation. SOPHIE I (Brown, Burton, & deKleer, 1982) for example,
used SPICE, an electronic simulator, as a black box to compute answers to
questions and evaluate student hypotheses. The WITS tutor uses a black-box
application interface to the live system. There are special issues that come
up when we must use this kind of interface.

Tracking the User.   Our assumption for WITS Tutor was that there is an
existing software application that could not be modified in any way, hence,
we had to treat the software application as a black box. However, by "in-
strumenting" the terminal emulator and user-interface toolkit, we were able
to capture each of the user's actions on the interface. This is similar to what
Miller, Hill, Masson, McKendree, Zaback, and Terveen (1987) did in BYGONE,
a component of the IDEA advising system. However, BYGONE did not have
to track usage of applications running a wide range of different user interfaces and hardware platforms.

WITS Tutor constantly monitored the student's inputs and matched them
against a task model. If the most recent input was deemed correct, it was
passed on to the software application to execute. If the input was deemed
incorrect, the tutor gave feedback and allowed the user to input again. In
some cases, the tutor erased the previous input so as not to confuse the
student.

Human tutors typically have the full history of the user's inputs, but is
this information used? If so, how? Hill (1993) put advisors into two conditions: a full-screen condition, where the advisors responded to user questions and had access to the user's history on the software application by
watching a display, and a no-screen condition, where the advisors responded without the benefit of a display. Forty-five percent of the replies
in the full-screen condition required knowing aspects of the prior interac-

tion, whereas 67% required knowledge of the current display. Clearly, history is important to generating good advice. However, not enough research has been done to determine precisely how history is used.

Levels of Granularity. What is an appropriate level of granularity for the tutor to track the user? The LISP Tutor (Farrell, Anderson, & Reiser, 1985) tracks users at the level of function names and arguments, and critic-based systems (Fischer, Lemke, Mastaglio, & Morch, 1991) evaluate entire solutions. Greer and McCalla (1991) have emphasized that tutoring needs to switch levels of granularity to improve both diagnosis and instruction. Providing different levels could be particularly important when tutoring users with a range of experience.

WITS Tutor utilized different levels of granularity for instruction but the tutor did not automatically adjust the level of granularity. First, our student model allowed for different levels of instruction, but the level could only be changed by the student. Second, in exercises we had two levels of granularity for problem solving: a decision-making level and a computer operations level. Once the user completed the problem solving level, they could move on to the computer operations level. Finally, for the computer operations, our user interface action language and procedure description language allowed for different levels of description. However, for the most part, the level of granularity was fixed. We decided that our character terminal users did not need help with typing and that our users with graphical interfaces did not need help at the level of mouse events like scrolling a text window or iconifying a window. We created a representation language that captured menu selections, command input, function-key pressing, and field editing at a symbolic level. This level of description was adequate for our needs.

Based on our experience with WITS, we recommend capturing the input at the lowest level possible and then providing a way of abstracting away the details so that the other modules of the tutoring system can work with higher level abstractions. The level of granularity should be controlled by the user indirectly by having him or her give feedback on the system's output. If the user does not understand some feedback, the system can break down the concepts into more specific steps or recommendations.

Recognizing Application State. What kind of interface to the live application software is necessary? An interface could be at the data layer, the processing layer, or the user interface layer of the application. To fully understand what the user is doing, a tutor needs to track both the user interface of the application and the student's mental state. A human tutor presumably infers the user's mental state from the state of the application, the user's history on the application, and from ongoing dialogue.

WITS Tutor tracked only the student's input to the software application, not the application state. Under ideal conditions, it could give proper responses, that is, if:

- The internal state of the software application (database and processing) was the same as when the task model was originally recorded.
- There were no external events that could impact the system.
- There were no unforeseen errors resulting from unexpected user data or system state.

If any of these assumptions were violated, WITS Tutor could become "out of sync" with the software application and could not "notice" that there was something wrong.

Synchronizing the Model and the Application. To combat the "out of sync" problem, we designed three enhancements to WITS Tutor:

- Set the initial state of the application.
- Do exercises in a strict order.
- Check the application state during exercises.

We set the initial state of the software application by starting a new session with the application and loading a training database. A new session meant that we had to either interrupt the user's current work, or we had to start a separate training window (which meant managing multiple terminal-emulator processes).

Because users could skip topics in the lesson plan, exercises could be done, potentially, in any order. One important enhancement for WITS Tutor was to add some control to exercise choice. However, enforcing a rigid exercise order seemed to undermine our goal of letting the users choose job-relevant parts of the course. Between the linear task model simplification and the exercise-ordering constraint, it seemed that we ended up with a "scenario machine" (Carroll & Kay, 1985). In reality, some exercises were prerequisite to others, whereas some could be done in any order. Also, some were optional, and others were required. We needed a more general way of handling prerequisites and ordering constraints.

Verifying Application State. Real tutors examine the screen, verify the effects of the student's input, and notice any errors or discrepancies. Although this may be a minor part of a tutor's skill, it is a major part of any expert user's knowledge of a software application. People learn what happens on the screen when they do various operations in a software application. User guides and new-user documentation spend a lot of time showing pictures of what the screen looks like when various operations are performed. Using a well-written user guide, a new user can detect their own errors by noticing discrepancies between the desired output and the actual output.

There are two types of state checking: error checking and result checking. Error checking usually involves noticing error messages on a message line or as pop-up warnings, or noticing that screen appearance is abnormal. We had no mechanism for tracking the appearance of the screens, and because there was some variability to the user's inputs and the system's processing, we could not simply compare a screen image with that generated by our task model. Also, each application's messages were specific to that application.

Result checking involves verifying that the software actually performs the actions that the task model says. WITS Tutor was able to "recognize" each of the screens using static labels found in particular positions on the screen and authored in the description editor. We used this ability to check where screen transitions were supposed to occur in the task model and then verify that the correct screen was reached. Unfortunately, if the correct screen was not reached, WITS Tutor was only able to give some general instructions and to wait until the student returned to the correct screen. We needed a more flexible system that could recognize an application state and plan a sequence of steps to cover from that state. This is a critical part of an expert's knowledge, but we neglected it in our design because we assumed that we could always keep the task model and the student in perfect synchronization.

Summary. In summary, WITS Tutor was an ambitious ITS targeted at industrial software applications. It was mainly concerned with the problem of tracking users on a live application, matching them to a task model, and giving useful feedback based on any mismatches. Based on our experience with WITS Tutor, we make the following recommendations to ITS designers, especially those designing intelligent tutors for commercial software applications:

- Carefully examine the tutoring system's context of use.
- Make the tutor secondary to the software system or simulation.
- Simplify task models to improve authorability.
- Use an integrated multimedia environment.
- Provide ways for students to volunteer information to populate the user model.
- Use different levels of granularity to increase the range of users that will find the system interesting.

- Do not ignore the recognition and recovery capabilities in the expert model.

PERFORMANCE SUPPORT

Now, more than ever, training is focused on improved productivity through human performance improvement. Having a performance orientation means thinking of all user support in the context of job performance. Here are the major changes for the training organization:

- *Integrated Support*—Help, documentation, and training are being prepared by the same organization. Information is being shared and reused by these organizations in creating learning-support information.
- *Performance Support*—Measurements of effectiveness are based on results on the job, not on measures of learning or satisfaction with training. Measures of speed and accuracy at job tasks are preferred to traditional measures of post tests (e.g., retention).
- *Just-Enough Training (JET)*—Most training goes unused because it was inappropriate to the individual's needs at the time. JET attempts to package training in small enough units so that just the right amount can be given to the student.
- *Just In-Time Training (JIT)*—Training should not be given before it is to be used. JIT attempts to connect the training with its use in such a way that training is given at the last possible moment.

There is a prevalent view in industry that ITSs are "next-generation CBT." This establishes a role for this technology in traditional training departments, but we believe that it is a misguided one that diminishes the potential impact that ITS can have on an organization. ITS should be viewed as one interactional paradigm that can be used to both facilitate learning and enhance performance on the job. In this view, the ITS is a form of electronic performance support and has the promise of wider interest and application in corporations.

Bellcore's Learning Support organization subsequently embarked on a number of performance support projects. One project is building task-oriented online help systems for Bellcore software applications using off-the-shelf authoring software. Another project is creating interactive multimedia reference materials for Bellcore software systems by recording expert users and adding tutorial narrations. Finally, the trainingless interface project aims to provide a more contextual, situated assistance for software application users.

Task-Oriented Help Systems

Our work on task-oriented help systems is aimed at providing task-completion advice, step by step procedures, and other performance aids in the context of the user's work on the application. The goals of this work are to:

- Reduce the time it takes for users to find and utilize help on user interface procedures.
- Provide training on user interface procedures online during task performance.
- Improve user satisfaction with the help system.

We have built a prototype for the NMA/EG (Network Monitoring and Analysis/ Enhanced Graphics) Bellcore product that we plan to field test in the near future at a customer site. Once the prototype has been tested, we plan to integrate it into the NMA product. The final NMA Task Help product will work as follows: When the user is in the middle of the application, they can select Task Help and a menu of tasks appears. The user selects a task and a procedure is displayed. The user selects a step and asks for an example. The system displays a textual description of how to do the step along with a graphic with the proper part of the screen highlighted. The user exits the help window, returns to NMA/EG, and completes their work, incorporating the information gleaned from the help system. Task-oriented help has had many advantages for the sponsoring software organization. It has:

- Made it easy for anyone to add task help information.
- Simplified the process of providing context to the performance support tool.
- Improved the help system features beyond their original text popup windows.
- Improved the appearance of the help system.
- Improved the scope of the help system.

The task-help software package consists of these five components:

- An application with a help access mechanism.
- A network communication package.
- A task-oriented graphical user interface (GUI).
- Application-specific data that specializes the GUI to the application.
- A code library that provides a task help API (Application Program Interface).

Our task-help package was implemented using the Icon Author™ authoring system. Procedures are represented graphically and buttons allow quick access to details about procedure steps. The fields or buttons being used in the step are highlighted in the graphic to promote the correspondence between the visual and verbal representations. The application-specific data added to the task-help package consists of graphics files that depict the software system in action and text files that specify the menu hierarchy, supply information about procedure steps, and give justifications for performing certain procedures.

The task-help package could form the basis for a performance-assistance interaction paradigm that could be reused. It also shows the profound effect that simplified authoring can have on a corporate training organization: The NMA trainers were able to quickly add large amounts of detailed information.

Interactive Multimedia Reference Materials

Project Peapod is an experimental effort to provide CD-ROM-based multimedia reference materials to users of some Bellcore products. Its target is a user who attended training or went through a tutorial but who needs to go back and take a quick refresher course before starting to use the system again.

Peapod is focused on step-by-step procedures for how to do tasks on computer systems. It provides a hierarchical menu system with procedure descriptions and narrated animations that show how these procedures are carried out on the software system. Narrated animations show abstract representations of how things work behind the scenes in the Bellcore systems. It includes "minimal manual"-style task steps (Carroll, Smith-Kerker, Ford, & Mazur-Rimetz, 1988) in text format for people who prefer a checklist-style approach.

The following circumstances led us to believe that the time was good for us to begin exploring multimedia as a means for providing performance support to our customers:

- In many cases, user desktop equipment was being upgraded from dumb terminals to moderate and high-end personal computers.
- CD-ROM drives were dropping in price.
- Several companies promised cross-platform, multimedia development environments.
- Portable CD-ROM players/viewers were being discussed by various vendors.

The first Peapod prototype was done in Hypercard™ using off-the-shelf multimedia tools. The current prototype is running in IconAuthor and uses Quicktime™ movies for all animations.

The initial reaction to Peapod was quite positive. User testing revealed that people tended to watch the animations and skip the textual descriptions of procedures, so we may create a future version with even less text.

The Trainingless Interface

The goal of the "trainingless interface" project is to make it possible for a user to sit down at an OSS with a large number of complex functions and begin to work immediately. We anticipate that this work will result in technology for building new user interfaces as well as for improving the usability of existing interfaces through performance support adjuncts.

Our first research goal is to demonstrate a performance improvement on usage of a software application by improving the online help system. Specifically, we want the help system to emulate the important aspects of one-on-one interaction that make it superior to the one-to-many situation found in classrooms and lecture halls (Bloom, 1976).

A significant aspect of one-to-one interaction is that it utilizes the user's context (the screen) and their history in the application. We conjecture that a help system that utilizes application context and interaction history can improve performance over a traditional help system. We will be able to side step many of the difficult issues alluded to in the section on the application interface because we are *retrieving* help information instead of providing specific advice or instructions for the user to follow.

Ultimately, we hope that this research will be able to isolate those aspects of one-on-one interaction that are particularly effective and design those into performance support tools such as help systems, "learn as you work" interfaces, and intelligent agents.

CONCLUSION

Recent trends in education and training philosophy are changing where, when, and how interactive learning environments interact with users. We are moving toward a model where people learn on the job or as they apply knowledge to real-world practice examples. Knowledge is no longer the province of the training organization but is distributed throughout a community of knowledge users. Thus, anyone is a potential author or trainee. Authoring tools to support this kind of community knowledge-sharing must be usable by a wider audience. We must also be aware that users of this knowledge have a context that drives their need for training and that context must be understood and articulated. We are now researching ways of making this kind of community knowledge sharing practical.

We believe that research should become more relevant by addressing its problems in the context of typical corporate environments; live equipment and software systems, large databases, real work environments, task variability, and individual differences can all impact how interactive learning environments are designed. Researchers should listen more to the needs and trends in corporate training: integrated, just-in-time, just enough to support performance.

ACKNOWLEDGMENTS

Thanks to all the members of the WITS team past and present including Tom Ackenhusen, Simon Blackwell, Doris Ip, Gupi Silverstein, John Smith, Ik Yoo, and many others; the WITS management (Adam Irgon, Margaret Nilson, Russell Sellars, Lynn Crilly); members of Bellcore's Applied Research area who have helped shape these ideas; Bernadette Cooney for support of the GSAT project; and Stacey Miltiades and Glenn Silverstein for starting it all.

TRADEMARKS

Framemaker is a registered trademark of Frame Technology, Inc.; Galaxy is a trademark of Visix corporation; Hypercard and Quicktime are trademarks of Apple Computer, Inc. IconAuthor is a registered trademark of AimTech corporation; Unix is a trademark of Novell, Inc.; Microsoft Windows is a registered trademark of Microsoft Corporation; WITS Author, WITS Tutor, PICS, and LEIS are trademarks of Bellcore.

REFERENCES

Anderson, J. R., & Reiser, B. J. (1985). *The LISP Tutor*. BYTE, April.

Anderson, J. R., Boyle, C. F., Corbett, A. T., & Lewis, M. W. (1990). Cognitive modelling and intelligent tutoring. *Artificial Intelligence, 42*, 7–49.

Atwood, M. E., Burns, B., Gray, W. D., Morch, A. I., Radlinsky, P. R., & Turner, A. (1991, June). The Grace Integrated Learning Environment—A Progress Report. *Proceedings of the Fourth International Conference on Industrial and Engineering Applications of Artificial Intelligence and Expert Systems (IEA-AIE91)*, pp. 741–745.

Bloom, B. S. (1976). *Human characteristics and school learning*. New York: McGraw-Hill.

Bonar, J., Cunningham, R., & Schultz, J. (1986). An object-oriented architecture for intelligent tutoring systems. *Proceedings of OOPSLA-86*.

Brown, J. S., Burton, R. R., & deKleer, J. (1982). Pedagogical, natural language and knowledge engineering techniques in SOPHIE I, II, and III. In D. Sleeman & J. S. Brown (Eds.), *Intelligent tutoring systems*. New York: Academic Press.

Carroll, J. M., Smith-Kerker, P. L., Ford, J. R., & Mazur-Rimetz, S. A. (1988). The minimal manual. *Human-Computer Interaction, 3,* 123–53.

Carroll, J. M., & Kay, D. S. (1985). Prompting, feedback, and error correction in the design of a scenario machine. *Proceedings of CHI 85: Human Factors in Computing Systems* (pp. 149–153). New York: ACM Press.

Clancey, W. J. (1987). *Knowledge-based tutoring: The GUIDON program.* Cambridge, MA: MIT Press.

Department of the Treasury, Internal Revenue Service. (1989). *Training 2000, Publication 1480,* Washington, DC.

Egan, D. E., Nakatani, L. H., Shaw, A. C., & Hawley, P. (1987). Empirical studies using TNT: Error diagnosis and coaching strategies. In J. Moonen & T. Plump (Eds.), *EURIT '86: Developments in educational software and courseware.* New York: Pergamon.

Farrell, R., Anderson, J. R., & Reiser, B. J. (1984, August). An interactive computer-based tutor for LISP. *Proceedings of the Fourth Annual Conference on Artificial Intelligence.* Cambridge, MA: MIT Press.

Fischer, G., Lemke, A. C., Mastaglio, T., & Morch, A. (1991). Critics: An emerging approach to knowledge-based human-computer interaction. *International Journal of Man-Machine Studies, 35,* 695–721.

Goldstein, I. P. (1982). The generic graph: A representation for the evolution of procedural knowledge. In D. Sleeman & J. S. Brown (Eds.), *Intelligent tutoring systems.* New York: Academic Press.

Gray, W. D., & Atwood, M. E. (1992). Transfer, adaptation, and use of intelligent tutoring technology: The case of Grace. In M. Farr & J. Psotka (Eds.), *Intelligent instruction by computer: Theory and practice.* Washington, DC: Taylor & Francis.

Greer, J., & McCalla, G. A. (1991). A computational framework for granularity and its application to educational diagnosis. *Proceedings of the 11th International Joint Conference on Artificial Intelligence* (pp. 477–482). Cambridge, MA: MIT Press.

Hill, W. C. (1993). A Wizard of Oz study of advice giving and following. *Human Computer Interaction, 8*(1).

Hollan, J. D., Hutchins, E. L., & Weitzman, L. (1984). STEAMER: An interactive inspectable simulation-based training system. *AI Magazine,* Vol. 2.

Lang, K. L., Graesser, A. C., Dumais, S. T., & Kilman, D. (1992). Question asking in human-computer interfaces. In T. W. Lauer, E. Peacock, & A. C. Graesser (Eds.), *Questions and information systems.* Hillsdale, NJ: Lawrence Erlbaum Associates.

Lefkowitz, L., Farrell, R., & Yoo, I. (1993). A knowledge-based intelligent tutoring system for telephone operations support systems. *Proceedings of the International Communications Conference.* Milan, Italy.

Macmillan, S. A., Emme, D., & Berkowitz, M. (1988). Instructional planners: Lessons learned. In J. Psotka, D. L. Massey, & S. A. Mutter (Eds.), *Intelligent tutoring systems: Lessons learned.* Hillsdale, NJ: Lawrence Erlbaum Associates.

Massey, D. L., de Bruin, J., & Roberts, B. (1988). A training system for system maintenance. In J. Psotka, D. L. Massey, & S. A. Mutter (Eds.), *Intelligent tutoring systems: Lessons learned.* Hillsdale, NJ: Lawrence Erlbaum Associates.

Merrill, M. D. (1989). An instructional design expert system. *Computer-based instruction, 16*(3), 95–101.

Miller, J. R., Hill, W. C., Masson, M., McKendree, J., Zaback, J., & Terveen, L. (1987, September). The role of system image in advising. In B. Shackel (Ed.), *INTERACT '87: Second Conference on Human-Computer Interaction.*

Muller, M. J. (1991). PICTIVE: An exploration in participatory design. In *Reaching Through Technology: CHI '91 Conference Proceedings* (pp. 225–231). New Orleans, LA: ACM Press.

Muller, M. J., Farrell, R., Cebulka, K. D., & Smith, J. G. (1992). Issues in the usability of time-varying multimedia. In M. M. Blattner & R. B. Dannenberg (Eds.), *Multimedia interface design.* ACM Press.

Murray, T., & Woolf, B. (1992). Results of encoding knowledge with tutor construction tools. *Proceedings of the Tenth National Conference on Artificial Intelligence.*

Murray, T., & Woolf, B. (1992, June). Tools for teacher participation in ITS design. *Proceedings of the Second International Conference on Intelligent Tutoring Systems (ITS-92)*, Montreal, Quebec.

Nwana, H. S. (1991). User modelling and user-adapted interaction in an intelligent tutoring system. *User modeling and user-adapted interaction* (pp. 1–32). Kluwer.

Polson, M. C., & Richardson, J. J. (1988). *Foundations of intelligent tutoring systems.* Hillsdale, NJ: Lawrence Erlbaum Associates.

Reiser, B. J., Anderson, J. R., & Farrell, R. G. (1985). Dynamic student modelling in an intelligent tutor for LISP programming. *Proceedings of the Ninth International Joint Conference on Artificial Intelligence.*

Russell, D., Moran, R. P., & Jordan, D. S. (1988). The instructional-design environment. In J. Psotka, D. L. Massey, & S. A. Mutter (Eds.), *Intelligent tutoring systems: Lessons learned.* Hillsdale, NJ: Lawrence Erlbaum Associates.

Sleeman, D. (1982). Inferring (mal) rules from pupil's protocols. *Proceedings of the European Conference on Artificial Intelligence*, pp. 160–164.

Sleeman, D., & Brown, J. S. (Eds.). (1982). *Intelligent tutoring systems.* New York: Academic Press.

Tarr, S. M. (1991, May). *Software Product Customer Satisfaction Survey.* Bellcore memo.

Towne, D. M., & Munro, A. (1989, September). *Tools for simulation-based training.* ONR Final Report No. 113.

Towne, D. M., & Munro, A. (1988). The intelligent maintenance training system. In J. Psotka, D. L. Massey, & S. A. Mutter (Eds.), *Intelligent tutoring systems: Lessons learned.* Hillsdale, NJ: Lawrence Erlbaum Associates.

7

TRANSFERRING LEARNING SYSTEMS TECHNOLOGY TO CORPORATE TRAINING ORGANIZATIONS: AN EXAMINATION OF ACCEPTANCE ISSUES

Peter T. Bullemer
System and Software Technologies
Honeywell Technology Center

Charles P. Bloom
Human/Computer Interaction Laboratory
NYNEX Science & Technology, Inc.

Over the past decade, significant technological advances in the areas of graphical user interfaces, animation, multimedia, artificial intelligence, and commercially available authoring environments have facilitated the development of advanced learning systems. However, despite these advances, widespread use of such systems has not yet been realized. From a corporate research and development (R&D) perspective, this chapter presents a case study on transferring learning systems technology to corporate training centers.[1] In this case study, we examine specific aspects of our experience that highlight pragmatic acceptability issues that are influencing widespread use of today's learning systems technologies, with the objective of examining the aspects of our approach that lead to the successful transfer of interactive learning technologies to our corporate training center.

---

[1]*Corporate* is used in the broader sense in reference to the corporate world as opposed to the educational world of schools and universities. Honeywell's corporate training centers and corporate research center are business units whose customers include internally owned corporate divisions as well as external corporate or government organizations. In the stricter sense of the term, many companies have corporate training or R&D organizations that provide training or research as a service to only their own business units.

This case study is based on an R&D project that started with an explicit research and development strategy to establish the credibility and feasibility of interactive learning system technologies with its corporate clients. The project led to the successful deployment of a multimedia, PC-based training application that gained immediate, widespread acceptance within the corporate user community, and the establishment of a program to continue the development and maintenance of similar systems within the corporate training center.

It is our premise that from gaining corporate sponsorship to managing projects to developing the applications, key acceptance issues must be addressed in order to achieve effective technology transfer. Moreover, within the corporate environment, an acceptable solution typically must satisfy the needs of multiple users and stakeholders. These key acceptability issues pertain to the nature of the proposed technical solution as well as to how the solution fits into the organizational fabric itself (Dede, 1995).

An important lesson learned from this case study is that sometimes, smaller steps in technology transfer can have a significant positive impact when the strategic approach emphasizes acceptability of the technical solution. Furthermore, several smaller steps may be necessary to achieve the long-term goals of widespread use of today's advanced learning systems technologies if the larger steps are met with resistance.

PRAGMATIC ACCEPTABILITY ISSUES

Corporate sponsorship is essential to obtain funding for initial technology transfer projects as well as for continued application development. The size of the technology gap—the difference between use of technology in current training practices relative to the proposed future practices—influences the ability to obtain acceptance of project proposals needed for corporate sponsorship. For example, a significant technology gap is present when the research center proposes to transfer intelligent tutoring systems (ITSs), and the current practices do not include the use of computers for instruction. A proposal to develop computer-assisted instruction to supplement existing classroom instruction creates less of an impression of a technology gap. The larger the technology gap, the greater the cultural changes implicated in accepting the proposed technology transfer (Hammer & Champy, 1993). A key to gaining corporate sponsorship is reducing the technology gap to an acceptable level of cultural change. In retrospect, we assert that the existence of this technology gap raised acceptability issues in four areas: demonstrating credibility, proving feasibility, justifying funding, and establishing corporate relevancy. As a result of an explicit strategy to reduce the technology gap represented in our project proposals, we were able to gain acceptance and corporate sponsorship.

Credibility

In general, the training center staff had significant doubts that a computer-based system could provide equivalent or better instruction. The computer-based approach was not perceived as leading to significant improvement in instruction. The current instructor-led approach emphasized hands-on experience, self-paced learning, modular design of instruction based on individual training needs, and instructor coaching based on observations.

The training centers were satisfied with their current approach involving classroom and hands-on training techniques. Computers were not used in the classroom for training purposes other than for demonstrating PC-based software tools. They were aware of computer-assisted instruction solutions but were not sure that they would provide added benefit to their current approach.

A demonstration of the potential of computer-aided instruction to teach concepts that were difficult to achieve through current classroom techniques led to an increase in credibility. Strategically, we proposed computer-based training as a way of supplementing the existing classroom approach as opposed to replacing it.

Feasibility

Proposed solutions that included the use of artificial intelligence techniques such as those used in ITS applications were rejected on the grounds of feasibility (Polson & Richardson, 1987; Psotka, Massey, & Mutter, 1988). Training center staff had significant doubts that authoring tools could be developed to enable their staff to build acceptable ITS solutions in a cost-effective manner. Awareness of the financial and human resource costs associated with knowledge engineering for expert system development influenced the perceived feasibility of the ITS solutions.

From a risk perspective, acceptable solutions were perceived to have low technical and business risk. Acceptable solutions were those that could be implemented using available, low-cost software development tools. Moreover, the solution could not require the end users to purchase equipment. Consequently, the target software and hardware platform had to be widely available.

FUNDING

These training organizations did not have the budget to fund the development of computer-aided instruction applications. The training center budgets are operational budgets designed to support instructors' teaching courses, supervising self-paced learning through direct interaction with products, and facilitating specialty workshops. Although it was proposed that computer-assisted instruction could potentially reduce training costs over the long term, the initial funding requirements of computer-based systems development were prohibitive. Consequently, we had to seek financial support in other locations within the corporate organization, such as marketing and engineering.

To get over the funding barrier, we proposed an initial project as a small feasibility demonstration study to identify a potential application area and the potential financial benefits. The modest initial financial commitment enabled us to establish credibility and demonstrate the potential benefits.

Relevance

In seeking corporate funding, the engineering organization was the most likely funding agency. The engineering organization has a program for sponsoring research and development in areas that might lead to new products or enhanced capabilities of existing products. Typically, highest priority is given to projects related to their major product lines. Although they perceived the interactive learning systems technologies to be of potential value, these technologies were not perceived as relevant to current product development.

To establish relevancy of this technology transfer effort, we had to present a strong business case in that we needed to show how the transfer of the technology could result in a profitable product or service for the training center as a business unit. With the collaboration of the training center and the marketing organization, we identified an application area that produced payback on investment within a one-year period as part of the initial feasibility study. The promise of an immediate economic impact helped to establish the relevancy of the project to multiple organizational stakeholders.

Conclusion

In conclusion, four pragmatic acceptability issues can influence the ability of a project proposal to gain corporate sponsorship. To begin work, a source of project funds needs to be committed. To obtain funding, the project needs to be relevant to the corporate business objectives. This issue becomes more complicated when the funding source (engineering, marketing) is not controlled by the business entity (training, field services) directly affected by the project deliverables. Beyond business relevance, the proposed solution must be perceived as feasible from both a business and technical risk perspective. Finally, credibility of the solution as a desirable alternative to current practices is necessary for the training organization to agree to participate. Consequently, our success in gaining corporate sponsorship was grounded in a proposal that was acceptable to both the funding agency (engineering organization) and the training organization. In addition, the project proposal represented a small technology transfer step relative to the existing state-of-the-art learning systems technology.

In the remainder of the chapter, we characterize the key components of our approach that enabled the development of a computer-assisted instruction (CAI) application that was able to satisfy the needs and requirements of all users and organizational stakeholders. In the next section, we discuss the project management approach that empowered successful completion of the project. From the project management perspective, the challenge was to establish and maintain commitment of multiple organizational stakeholders. In the following section, we characterize the technical approach that enabled us to build an application that was acceptable to the student users as well as the multiple stakeholders. In the final section, we highlight the significant positive impact of this initial technology transfer project on the corporate training environment and reflect on the potential for transferring more advanced interactive learning technologies.

MANAGING MULTIPLE COMMITMENTS

A crucial element of our project management approach that enabled successful completion of the project was the establishment and maintenance of commitment of multiple organizational stakeholders. The strength of commitment was directly related to the degree of project acceptance by these stakeholders. Throughout the project, the extent to which specific needs of individual stakeholders were met influenced their degree of acceptance of the project. In this section, we present our perspective on the stakeholders' needs and how the project management approach influenced our ability to understand and accommodate them.

Understand the Diverse Needs of Multiple Stakeholders

In the previous section, we identified four organizational elements involved in establishing this project: marketing, engineering, training, and research. As stakeholders in the project, key personnel within these organizational elements had a specific interest and desire to gain from this project.

Marketing. The main interest of the marketing organization was to help the field service organization reduce their operating costs and improve their financial performance. The proposed CAI application provided an opportunity to reduce operating costs through a reduction in training costs. The amount of benefit was directly related to the extent to which the CAI application produced an increase in "time on the job" and a decrease in the costs associated with the travel and living expenses incurred while attending class at the training center.

Consequently, project decisions that impacted the size of the benefit affected marketing's level of commitment. For example, a decision that resulted in a reduction of the curriculum breadth reduced their degree of acceptance. For the marketing organization, pragmatic results were more important than proving the learning efficacy of the application or enhancing the technological sophistication of the solution. These individuals were interested in customer (i.e., the field service organization) satisfaction and hence, generating the perception of meeting customer needs.

Engineering. The main interest of engineering was to successfully transfer a new technology to a corporate business unit (i.e., the training organization) in the form of a prototype product. The CAI application development provided an opportunity to transfer state-of-the-art multimedia interactive learning technologies. The amount of benefit to the engineering organization was directly related to the quality of the technical solution and the extent to which the CAI solution embodied the latest in interactive, multimedia capabilities.

Obviously, project decisions that increased the technical quality, sophistication of the solution, and breadth of the multimedia interactive features resulted in greater engineering acceptance. For instance, the engineering organization was willing to accept a reduction in curriculum breadth if the amount of interactivity was increased through more animation and simulation. In terms of quality, demonstrating the learning effectiveness of the CAI application as well as developing bug-free software was important. For this research project, engineering was less interested in customer satisfaction and the reduction in field service's operating costs than in a demonstration of CAI technologies and proving their efficacy.

Training. The primary interest of the training organization was to meet the training needs of the field service organization through the delivery of high-quality instruction in a timely and cost-effective manner. The CAI development provided the opportunity to improve the timely delivery of training as well as to free up instructional staff to devote their time to customized courses driven by specific customer training requests (a large part of the training organization's "business"). The training center's instructional staff had a high course-delivery workload that limited their ability to introduce customized courses to meet one-time training needs. The introduction of a CAI application to supplement classroom training provided them with more flexibility to develop and conduct just-in-time, customized instruction. The amount of benefit to the training organization was directly related to the extent that the CAI application reduced the instructor's time in the traditional classroom. Furthermore, the reduction in the instructor's time in the classroom depended on the degree to which the instructors and the customers (field service students) perceived the use of the CAI as an acceptable substitution.

Project decisions that impacted the quality of the instructional content, breadth of curriculum coverage, and the satisfaction of the student users affected the training organization's acceptance. There was a need to demonstrate to the training organization that the CAI application was as good as the current instruction. Hence, the training organization was less interested in proving the learning efficacy of the CAI application than in determining the amount of customer (field service student) satisfaction.

Another aspect of the instructor workload situation that affected the training organization's acceptance of the project was the requirements imposed on the training center staff to serve as subject-matter experts to the project. Instructors did not have extra time to devote to supporting the CAI development. In many cases, instructors donated personal time to support development activities. Although the development team was extremely flexible and adapted their interactions with instructors to minimize the impact of development on the instructors' preparation and classroom time, the instructors did not immediately show commitment to the project. However, once the instructors saw that their needs were being met, they were more likely to make time to assist in the development activities.

Research. The primary interest of the research organization was to enhance the instructional delivery capabilities of the training organization through the transfer of interactive learning technologies. This project was perceived as an initial step to cultivate opportunities to transfer additional, more advanced technical solutions. The CAI application development provided the opportunity to establish the credibility and feasibility of interactive learning technologies. The strength of the benefit to the research organization was directly related to the extent to which the CAI application was perceived as an acceptable substitution for the current method of instruction as well as a technical solution that the training organization could develop and maintain within its organization.

Project decisions that added risk in delivering a CAI application that was acceptable to the end users (i.e., instructors and students) affected the degree of acceptance by the research organization. For example, extending the curriculum breadth at the cost of adding interactivity to improve user acceptance reduced acceptance by the research organization. For the research organization, delivering a working CAI application that provided a sufficient level of interactivity to keep the student's interest was more important than maximizing the level of technical sophistication or use of state-of-the-art multimedia technologies.

In conclusion, the diversity of needs of the multiple stakeholders presented a significant challenge to the success of the project. Whereas understanding and adjusting to these complex and potentially conflicting needs is critical, the discovery and recognition of stakeholders' needs is not a straightforward or simple endeavor. Often, these tacit needs are revealed at critical decision-making junctures in the project, such as design reviews or project reviews, when strategic decisions influence the project direction and scope.

Adapt to Emerging Constraints

The project management approach influenced our ability to understand and accommodate the specific needs of the four organizational elements to maintain commitment through the duration of the project. Although ideally we attempted to maintain the commitment of all stakeholders throughout the project, in practice this was not always possible. More importantly, the need for commitment from different stakeholders depends on their specific interests, their sponsorship role, and the development stage of the project. In the remainder of this section, we describe the key components of the approach and their role in maintaining sufficient commitment to meet the project objectives.

Weekly Project Status Meetings. The core application development team, consisting of three individuals within the research organization, held weekly status meetings to assess progress against goals and to evaluate our technical approach on a regular basis. We were continually adapting our technical approach to emerging constraints imposed by the organizational stakeholders.

More specifically, we continually adapted our knowledge-acquisition method to the kind of information available, the accessibility of information sources, and the interaction preferences of the individual instructors. Moreover, unexpected changes in availability of subject-matter experts or materials forced adjustments to schedules and objectives. Later, weekly contingency plans were developed to empower developers to make adjustments as the situation demanded them.

Throughout the project, these status meetings were used to prioritize development goals. Each week, a list of prioritized development activities was generated, based on the application module development schedule, that identified software bugs and recommended improvements from design reviews and student evaluations. The priority scheme was based on perceived impact on the application acceptability to users and stakeholders. This approach to managing development activities enabled us to meet a fairly ambitious development schedule.

Managing the Client. In most projects of this nature, individuals outside the core development team get involved in managing the project. These stakeholders, as clients, are looking out for their special needs and interests. Without preparing for this interaction or establishing explicit roles and responsibilities, there is the likelihood of a negative impact on project progress. For instance, the engineering organization assigned a project manager to supervise the project. This project manager demanded to review implemented application modules on a weekly basis. We found ourselves constantly doing extra work, preparing for his reviews, to justify our design decisions and to elucidate implementation areas still under development. The lack of understanding and agreement on the approach led to a general dissatisfaction with the team's progress and a reduced commitment from the engineering organization.

To help us manage our "client," we instituted periodic management reviews during which input on modules and software issues was directly solicited. Further, to keep our clients informed (but not interfering) on a weekly basis, we distributed the issue and activity list, generated in our weekly review meetings, to communicate development goals and progress to key individuals in the training and engineering organizations. This communication improved the general understanding of the basis for the weekly development goals and, subsequently, reduced the need for additional explanation and justification.

Project Coordinator at the Training Center. In the first phase of the project, a single instructor was assigned to support the project. The limited availability of this instructor impacted the developer's subject-matter acquisition. This instructor had a full-time commitment to conduct classroom instruction. In order to support the project, the instructor had to put in additional hours, an arrangement that had a significant impact on the development schedule.

In response to this issue, we worked with the training center management to identify an individual to be a project coordinator. The project coordinator provided a single point of contact at the training center to facilitate access to several instructors, subject-matter experts, and students. The project coordinator's knowledge of the overall capabilities of the training center staff and classroom schedules enabled timely and appropriate access to resources. As a result, the project was able to take advantage of a greater diversity of knowledge and materials in the training center.

Periodic Management Review Meetings. The most effective mechanism for interacting with management across the organization was in periodic meetings with representatives of each organization. As mentioned earlier, each of the organizational stakeholders had specific needs to be met that influenced its participation and commitment to this project. The management review meetings were the primary forum for expressing those needs and allowing each stakeholder to influence the direction of the project. Each management review meeting involved consensus decision making on strategic issues pertinent to the current project stage. The critical issues changed at different stages of the project with respect to maintaining management commitment. Over the 12 months of the project, we conducted three management review meetings.

In the start-up stage of the project, the meeting objectives were to present the results of the feasibility analysis, gain consensus on the application scope, and obtain support for the project development plan. At this meeting, each of the organizational stakeholders had an opportunity to influence the project objectives. Obtaining consensus at this point was important to getting organizational commitment of resources. In particular, the commitment of the training (i.e., instructors, subject-matter experts, and students) and engineering (i.e., funding) organizations was most critical at this juncture.

Following the initial design and implementation of the first instructional module, the meeting objectives were to gain acceptance of the CAI design and the project development plan. At the initial meeting, the management agreed to an abstract, functional description of the CAI contents. With the demonstration of a working prototype system, the management stakeholders had a concrete understanding of the project deliverable. Moreover, the discussions that followed the demonstration revealed that individuals differed in their expectations of the project deliverable. During these discussions, we identified and prioritized alternative or additional functionality and features as well as issues to be evaluated during testing. The ability to influence the design in this early stage strengthened the management commitment to the project. The increased enthusiasm and allocation of training center resources during this meeting were indications of increased commitment. Training organization commitment was most critical at this juncture because of the increasing demands for instructor and subject-matter experts' participation.

At the midpoint of the application development, the meeting objectives were to provide instructor and student evaluation results and set expectations for the final project deliverable. When the project was started, we identified minimum and stretch goals. At this point, the development methodology and technical approach were fairly well established. Consequently, estimations of the likelihood of achieving the project goals were more reliable. Each management representative was provided with an updated prototype one week before the meeting. This gave each of them an opportunity to use the CAI application and develop a firsthand understanding of the breadth and depth of the prototype. In particular, the commitment of the marketing and engineering organizations was most critical at this juncture.

Strong acceptance of the CAI prototype was obtained through the combination of direct experience with the prototype by the management representatives and results of the student evaluation at the training center. This strong acceptance increased the commitment of the training and marketing organizations. This commitment manifested itself in advocacy to complete the project and extend the project funding to ensure achievement of a broader CAI curriculum scope.

The engineering organization acceptance level was not strong enough to maintain their commitment to the project. The prototype was not meeting their needs in terms of interest in advanced, multimedia, interactive learning technologies. Moreover, the project had moderate priority for this organizational stakeholder due to relevancy to their product development. Consequently, the engineering organization withdrew funding due to low commitment and pressure to invest in another project perceived to be more relevant.

In conclusion, the project management review meetings were critical to maintaining the commitment of the organizational stakeholders through the first half of the project. The meetings were structured to enable all organizational stakeholders to participate in strategic planning and decision making, enabling the project team to adapt to emerging constraints to improve project acceptance. The resulting strengthening of commitment of the organizational elements directly affected by the project outcomes enabled the project to continue, despite withdrawal support from the funding agency, the engineering organization. Once a strong level of commitment was established, organizational representatives directly benefiting from the project results expended extraordinary effort to remove barriers to success, such as financial sponsorship. In general, as organizational commitments increased with acceptance, problems were more easily resolved through collaborative efforts because the organizational priorities were high.

DEVELOPING FOR ACCEPTABILITY

Another contributing factor to the success of this project was the technical approach. Certain aspects of the technical approach were modified during the development phase in response to problems that threatened achievement of project goals. These modifications highlight development process issues that impact system acceptability. The technical approach consisted of three stages:

- Define the application scope and expected benefits.
- Develop the application in iterative, incremental prototyping cycles.
- Transfer technical capability to corporate training center.

Key components of our technical approach included a domain feasibility analysis, participatory design process, incremental software development, modular system architecture, usability evaluation, and a commonly available multimedia hardware/software platform. In this section, we discuss the stages of the technical approach in terms of key components that led to successful deployment of the interactive learning system.

Define the Application Scope and Expected Benefits

The first stage of the development process involved defining the application–development objectives in terms of specific deliverables that could be met within the project budget and schedule. The overall project objective, as mentioned earlier, created a situation in which we were looking for a problem to meet a specific kind of solution. Moreover, for this particular project, the type of solution that was feasible was constrained by the project budget and schedule. Consequently, our initial constraints were the form of the solution (i.e., computer-assisted instruction as opposed to intelligent tutoring), a one-person-year level of effort, and a 9-month time frame. Through the process of conducting domain and feasibility analyses, we identified additional constraints that further defined the acceptable solution space.

As a first step, we conducted an analysis of courses at the corporate training center to identify a course that was appropriate for a CAI application. Moreover, because of the strong organizational need to succeed, we looked for an application that held the greatest potential for significant impact. Consequently, the initial list of candidate courses was generated by identifying those with the greatest annual enrollment. In addition to total volume of students, we assessed the continued need for training in the current subject matter and the stability of the subject matter.

In the second step of the analysis, we examined the availability and accessibility of instructional and course content knowledge sources (i.e., instructors, subject-matter experts, and course materials). Information on the availability and accessibility of knowledge sources indicated the relative level of effort required to develop the instructional content and strategy.

In the final step, we calculated the potential payback on investment. Calculating the benefits of training in terms of payback on investment is a conservative way to measure positive impact. Many of the positive benefits are difficult to measure directly. However, in meeting the needs of the management stakeholders, we were required to show some tangible results. We chose to measure benefits in terms of reduced cost of training. With the CAI application available in the workplace or at home, we could show a reduction in travel and living expenses (T&LE) associated with students attending the course at the corporate training center. Another potential source of reducing the cost of training is reducing training hours. However, we did not use this in our calculations because we did not have a basis of estimating it from past experience. As it turned out, the use of T&LE alone produced a payback on investment within 9 to 12 months. This payback was significant enough to obtain strong initial commitment from each of the organizational elements.

As a result of the domain analysis process, we narrowed the focus to two introductory courses that had the largest volume of students and the greatest stability in course content. The two introductory courses taught similar concepts to two different elements of the field organization—sales and operations.

Of the two selected introductory courses, the operations course-training objectives and instructional requirements were not well documented, with some contained in instructors' notes and others in the instructors' heads. The sales course objectives and instructional requirements were better documented but still incomplete. In addition, the training materials for both courses were distributed throughout a large volume of reference materials. These issues indicated that some effort was needed in the areas of knowledge acquisition and organization prior to system development for these courses. In addition, we found that not all aspects of these two courses were appropriate for the CAI solution, such as hands-on lab and field experiences. Furthermore, each course had more content than could be implemented within the budget and schedule of the project.

As a result of the analysis, we modified our initial strategy of developing a solution for one course to that of developing a CAI application to cover the course content that was common to the two related courses. Our revised strategy enabled us to develop one CAI application to partially meet the curriculum scope of the two courses with the greatest student volume.

The project-scoping stage culminated with a management review meeting. In the management review meeting, a recommendation for the project objective was presented, based on the domain feasibility analysis. Management accepted the recommendation to develop a prototype CAI application for the portions of the corporate training center introductory course that taught the basic product overviews to branch sales and operations' new hires. As a result of the review meeting discussion, additional scope constraints were added to insure that the project made an acceptable impact.

The benefit to the project depended on the reduction of time in the introductory course at the corporate training center. This critical factor led us to focus on two aspects of the project scope: curriculum coverage and target delivery platform. In terms of curriculum coverage, a goal was set to eliminate 3 to 5 days of instructional time at the corporate training center by enabling students to train at their remote branch locations prior to coming to the corporate training center to receive the remaining training through the traditional classroom method. In terms of target delivery platform, the ability of employees (as students) to use the CAI application in the branch office setting or at home was critical to the success of the project. Consequently, we decided to develop the application to run on the most commonly available hardware/software platform. The target platform was an 80386 class PC with 14" VGA color monitor, 4 Mbytes RAM, and Windows (3.0 or above).

In summary, the objective of the project was to develop a CAI application for the portions of two introductory courses that covered basic control fundamentals and product overviews. As a research and development project developing a working prototype, we defined minimum and stretch goals to establish stakeholders' expectations for the potential range in project results given the perceived areas of project risk. Our minimum goal was to produce a 30% (3-day) reduction to both sales and operations basic courses (10-day course). Our stretch goal was to produce a 50% (5-day) reduction. The reduction would result from new hires in sales and operations using the CAI application prior to attending the basic course at the corporate training center. Finally, the CAI application would run on a commonly available software and hardware platform.

Develop the Application

Once the scope of the CAI application had been specified, the next stage was to implement the application using an iterative prototyping approach. This approach is very similar to the approach of Successive Approximation (Allen, 1992) used in courseware product development. The steps in successive approximation process are:

1. Do content analysis, get to the mouse early, and prototype in delivery medium.
2. Get reactions.
3. Keep what works, refine what almost works, and build what is still needed.

Our emphasis was to work in the delivery medium, that is, to rapid prototype the CAI on the identified hardware/software platform from the onset of the project. This approach results in one common representation continually available that supports all participants, from designers to instructors to students to management, constructively participating in the development process. Such an approach was necessary as we had stakeholders reviewing the project continually. At any point or at every point in the project, there was a CAI prototype for evaluation. In contrast, in a traditional systems design approach, there were "representations" of the CAI available for review at every point in the project, but these "representations" were in different forms to which only a select few could react. Functional specifications, implementation plans, parts of the implementation, and test plans are examples of abstract, traditional systems design representations that can often be difficult for stakeholders to effectively review. In addition, having a "running" prototype always available for review enabled us to address usability issues beginning in the earliest stages of development.

Incremental Modular Development. In our analysis of the course content, we identified specific content areas corresponding to distinct topics taught in 4- to 8-hour segments. To achieve the goal of 5 days' coverage of the existing courses, we needed to implement course material for 10 topic areas. We developed an incremental implementation plan corresponding to successive, iterative prototyping cycles for each topic area. In the incremental approach, the steps of the successive approximation process are repeated in each prototyping cycle. Each cycle was approximately 4 weeks in duration. The software developed for each topic area was a separate software module. The incremental modular development of the course content enabled the development team quickly to evaluate the usability of the design, assess the accuracy and effectiveness of the course content, and identify bugs in the software architecture.

Staying within the development schedule was important to the success of the project. In order to meet the minimum coverage of the course curriculum, we created a fairly aggressive schedule. In implementing the course materials, the development team could easily go into a lot of detail on one topic area and quickly run out of time to cover all the topic areas. Particularly, the development of animation consumed much of the development time (Kodali, 1994). We found ourselves constantly making trade-offs in terms of how best to present the lesson material given the constraints of the schedule.

After recognizing the impact of implementing animation on our ability to deliver on schedule, we adopted a successive refinement strategy within each prototype cycle. We completed the implementation of all the graphical presentations as static graphics before we gave them dynamic properties through animation. We prioritized the importance of animation to understanding the concepts to be learned, based on the developers' intuitions. These priorities guided the implementation of the animation. Before we released a lesson for usability testing, we completed the high-priority animation. The usability evaluation data provided guidance to the development team of areas that might improve with animation or might need revision to implemented animation.

An additional advantage of the incremental modular approach was the early demonstration of the CAI's instructional potential. After each topic area was implemented and the initial usability evaluation was completed, the instructors continued to use the CAI in the corporate classroom setting to replace the previous lecture-based instruction. The instructors found that the interactive features of the CAI provided a better medium for teaching certain concepts than did their previous methods. This early demonstration strengthened the commitment of the training center staff and motivated them to increase their level of participation in the development process.

Participatory Design Process. From the beginning, we tried to involve many different kinds of individuals in the design process. A participatory design process typically involves the use of a pool of representative users to evaluate the application design (Nielsen, 1993). In our project, the users consisted of students attending the introductory course at the corporate training center and the course instructors.

From the perspective of curriculum design, the instructors were a valuable source of information. Initially, we were able to gain participation only at the end of the prototyping cycle. As the commitment and level of involvement of the training center staff increased, we were able to involve instructors in the earlier design phases. The instructors began to participate in brainstorming sessions, paper mockup critiques, and prototype reviews. This early involvement reduced the amount of revision required following the usability evaluations with students. Students were used throughout the project only at the end of the prototyping cycle. The student participation is discussed in more detail later in the context of usability evaluation.

From the perspective of the user–computer interaction design, we obtained frequent participation from user-interface design specialists within the corporate research center in the first half of the project schedule. These individuals provided critique of the user-interface conventions and interaction style from their understanding of "best" practices and human-factors standards. In addition, these conventions were evaluated for consistent application in each prototyping cycle. Following the heuristic evaluations of the specialists, we obtained empirical data from the usability evaluations conducted with students.

Sequential Versus Concurrent Design Process. In the first half of the project, we adopted a sequential design process involving the development of one topic area at a time. One individual acted as the lead designer and generated an initial design in the form of paper sketches. As a group, the development team members reviewed and revised the initial design sketches. The design and review activity was continued as different development team members implemented distinct parts of the individual lessons comprising the topic area. The initial design sketches identified only the high-level concepts to illustrate and the types of images and animations to use. Our emphasis was to get to the computer medium quickly and complete the design where the designer can deal directly with many of the design constraints that are imposed by the medium itself.

At the midpoint of the project, we had slipped significantly from the original schedule. We were on target to deliver the minimum goal of 3 days or six topic areas. This slippage was due in part to the limited participation of the training center staff in the initial activities of the prototype cycle, the amount of time spent on revising the user-interface design for usability, and the level of effort devoted to implementing and fine-tuning animation.

In the midproject management review meeting, a couple of participants emphatically suggested that project success depended on achieving the stretch goal for curriculum coverage. Only the 5-day coverage enabled the training center to reduce the course from 2 weeks to one week. This revised project goal forced the development team to reevaluate the development approach to find a way to improve productivity.

To meet the revised objective, we had either to shorten the prototype cycle time or develop topic areas concurrently. Shortening the prototype cycle time was clearly not feasible, given the limited availability of instructors, subject-matter experts, and students. Consequently, we decided to take a concurrent design approach.

Each individual on the development team was capable of developing a lesson from scratch (i.e., develop lesson objectives; obtain course material; develop graphics, animation, and text; conduct usability assessments; and modify based on the results). We found that concurrent development of topics optimized our ability to adjust our schedules to the availability of instructors, subject-matter experts, and students. At the same time, we improved our capability by enabling any team member to help another to meet schedule demands. At a given time, a team member had responsibility for several lessons in different stages of development.

In the first half of the CAI development, as much as 70% of the effort involved design activities. In the second half of the project, most of the effort focused on the development of course content and only a small portion on the design. This shift in level of effort was due in part to the stabilization of the application design. In particular, the user–computer interaction design was changed significantly in the initial three prototyping cycles (i.e., corresponding to topics). When the design conventions were fairly stable, we altered our design process from a sequential process to a concurrent design process. It is questionable that the concurrent approach would have worked as effectively prior to stabilizing the user–computer interaction design. However, we might have benefited from switching to the concurrent approach after the second prototyping cycle.

Knowledge Acquisition. The existing introductory-course training objectives and instructional requirements were not well documented, with some being contained in instructors' notes, and others in the instructors' heads.

In development of many of the topic areas, particularly in later development cycles, the instructors were designing new lessons that enhanced and extended the scope of the existing course. After observing the potential of the computer-based medium, the instructors found they were able to present concepts that were difficult to present with typical classroom materials. As a result of these factors, knowledge-acquisition activities were critical to the success of the CAI application development process.

The knowledge-acquisition approach might be characterized as opportunistic. We used whatever resources were available from existing course materials, instructors' notes, instructors filling out forms, interviews, and classroom observation. In this particular environment, we found the most effective approaches involved:

• Define the learning objectives—An instructor was asked to provide the learning objectives for the topic area. In addition, the instructor prioritized the objectives in terms of need (high priority), want (medium priority), and wish (low priority). After observing the "extending scope" phenomenon, we added the assessment of priorities to help contain the scope of topic coverage and make trade-off decisions in meeting the project schedule.

• Study available materials—The developers studied all available instructional or subject-matter materials. This activity usually began with one member of the development team attending the current class at the training center.

• Conduct brainstorming session on instructional approach—A lead designer of the development team and an instructor generated a set of lessons that met the learning objectives. For each lesson, specific concepts to be learned were identified. For each concept, the alternative instructional techniques were suggested.

• Review the design in paper sketch form with instructor—The development team and the instructor reviewed the initial design as a group.

• Fill in form for descriptive text and test items—Depending on availability, the instructor and/or a member of the development team created the descriptive text and test items for each of the lessons. A template form was developed to facilitate text development by the instructors.

Usability Evaluation. The best data on usability came from studies with students at the training center. As mentioned earlier, we made extensive changes to the user–computer interaction framework as well as to the course material itself in the initial prototyping cycles. Many of the animations had as many as 12 iterations. In the final prototyping cycle, only a few changes were made following the student evaluation, and these were primarily revisions to errors in the descriptive text.

We had the fortunate opportunity to use the students attending the class at the corporate training center to evaluate the usability. During the period of development, 12 classes participated in the evaluation. Approximately every other week, a new class was available. As each topic module was implemented, the module was integrated into the prototype and used in the classroom setting. Consequently, our pool of user representatives was continuously changing. As we revised a module based on the usability data, we were able to obtain new data from a new set of students.

At first, the instructors did not trust that the CAI application was effective. In fact, they imposed constraints on the use of the application in the classroom that reduced the value of the usability data. In particular, the instructional effectiveness of the application was difficult to assess because the instructors insisted on first teaching the course and then letting the students use the CAI application. Despite this constraint, we were able to get valuable usability data.

After the second prototype cycle, we persuaded the instructors to start with the CAI application and then follow up with other classroom techniques to make sure students achieved their learning objectives. Without the prior exposure to the course content, new usability issues were identified. Consequently, significant improvements in the curriculum structure and user–computer interaction framework were enabled by the unbiased interaction.

In the final stages of the project, the instructors had developed a significant level of trust and confidence in the instructional approach. As a result, the application modules that had gone through a revision following the initial prototype cycle were used as a substitute for the previous course instruction. The instructors used the remaining time to collect evaluation data and make recommendations for application revisions.

Transfer the Technical Capability

Before the CAI application had been completely implemented, we began steps to transfer the technical capability to update, extend, and maintain the application. This final stage of the technical approach was critical to achieving system acceptability from the perspective of application ownership. Acceptability of ownership meant that the application was perceived to be a product of the corporate training center rather than of the corporate R&D center. If the training center accepted ownership, we believed they were committed to keeping the application in use through maintenance activities such as keeping information current and fixing software bugs. In addition, the transfer of the technical capability enabled the training center to develop additional CAI applications.

The approach to technical capability transfer had two components:

- Establish a mentoring relationship between the development team and a member of the training staff.
- Document the system architecture design and the process for authoring application lessons.

A key aspect of our approach was to begin this process during the application development itself. The transition of the capability occurred more as a natural course of events as opposed to something that just happened at the end of the project. In our experience, many past failures in technology transfer can be attributed to lack of ownership of the technology and lack of technical capability to continue the technology development.

Mentoring. The objective of the mentoring activities was to work closely with an individual within the corporate training environment to develop the knowledge and skill to modify the content of any CAI application module and to extend the application with new modules. We sought an individual who was an instructor in the areas relevant to the CAI application curriculum and had some vested interest in the use of the application in the corporate training environment. In collaboration with training center management, we selected an instructor who participated in the module development activities and expressed interest in learning how to do the software implementation. Our development team worked closely with this individual in designing the final application modules. As these modules were evaluated for bugs and usability issues, the individual worked with the development team to revise the underlying software. We continued this mentoring relationship for the first year after the application was distributed as a learning product.

Documentation. The document used to support the transfer of technical capability was the software maintenance document. This document contained the system architecture design, authoring notes, authoring templates, user comments, and a prioritized bug list. The software maintenance document was a living document that was updated when design changes were made and when reviews were completed.

The software maintenance document was initially created to communicate the software architectural design among development team members involved in the implementation activities. The system architecture design described the functional components and associated features. It also illustrated how the functionality was embodied in the application infrastructure.

The authoring notes illustrated how the software tools were used to implement the application infrastructure. In addition, the authoring notes provided design rationale for the infrastructure. The authoring templates were designed to facilitate the knowledge-acquisition activities associated with the design of a single instructional module. The format of these templates mapped directly onto key functional components of the infrastructure to facilitate the implementation tasks.

The user comments and prioritized bug lists provided a record of design issues that were resolved or not resolved. This information was valuable for communicating usability design concerns as well as for tracking progress in refining the application.

The software maintenance document served multiple purposes. Initially, the document supported communication among development team members. As the project progressed, it facilitated transfer of knowledge to new members to the development team. Over the course of the 12-month project, we had four new members on the development team. Finally, this document was used to transfer the "know-how" to the corporate training center.

In conclusion, the transfer of the technical capability was an important component of the overall technical approach. The successful transfer of the technical capability, in addition to the project management and technical approach, had significant impact on expanding the use of CAI application in the corporate training environment.

EVALUATING THE TECHNOLOGY TRANSFER STRATEGY

This project started with an explicit research and development strategy to establish the credibility and feasibility of interactive learning system technologies with its corporate "clients." The long-term objective was to obtain support for more advanced research and development of intelligent interactive learning systems. The initiative led to the successful deployment of a multimedia, PC-based training application and the establishment of a project to continue the development and maintenance of similar systems in the corporate training center. The firsthand experience in the development and use of the computer-aided instructional applications has resulted in a perceived need for more advanced interactive learning system capabilities. Although the legitimacy of computer-aided instruction has paved the way for transferring more advanced learning systems technology, significant pragmatic barriers must be addressed before the transfer of advanced intelligent interactive learning systems becomes commonplace.

Significant Positive Impact

The project had a significant, positive impact on the current delivery of training in the corporate training center, as well as an altered perception of future needs for interactive learning systems. In terms of the R&D goals of establishing the credibility and feasibility of interactive learning systems, the project was successful in several ways.

In the beginning of this project, individuals at different levels in the corporate training center had doubts concerning the relative benefits of computer-assisted instruction. A year and a half after the completion of this project, the corporate training center had nine ongoing CAI projects. From the customers of the training center to the top management, the legitimacy of computer-assisted instruction is established.

The success of this technology transfer project was reflected in a significant customer demand for more CAI applications. The widespread distribution of the application resulted in broad visibility across the organization. This visibility was supported with positive evaluations of the CAI capabilities.

In addition to the positive impact on demand for CAI applications, the project provided the corporate training center with a strong foundation on how to develop CAI applications. Although the corporate training center has begun building an internal development team, they have sought support from third-party training, software development companies, to meet the current demand. Importantly, the involvement of the corporate training staff in our project has provided them with critical knowledge on how to manage the CAI development projects. In particular, they have continued to use the project coordinator role to facilitate the interaction of the development teams with instructors and subject-matter experts. Moreover, the project coordinator has demanded changes to the third-party developers' approach to development, including participatory design and usability evaluation.

Along with the increased use of CAI applications and the experience in deploying this type of training, the corporate training center has developed a perceived need for advanced capabilities in both the distribution of CAI applications and the delivery of the instruction.

With the advent of multiple forms of training available on the corporate network, employees are confronted with the difficult tasks of finding what is available, assessing whether it applies to their training needs, and developing a training plan that is compatible with individual training budgets. To enhance the potential economic benefits derived from CAI applications, the corporate training center will need to offset the potential negative impact of searching for the appropriate applications from those available on the network. Consequently, they perceive a need for an online computer-managed instruction (CMI) system that can assist in developing a training curriculum based on individual needs, track employees' performance, document core competencies of the workforce, and notify individuals of need for retraining based on certification requirements and new product releases.

In addition to the potential negative impact of locating appropriate instruction, the extended use of CAI applications has led to a perceived need for more adaptable CAI delivery. Through the process of developing additional CAI applications, the corporate training center realizes the difficulty in designing a single CAI curriculum that adequately meets the needs of all users. From a user acceptance perspective, the CAI customers are concerned with the cost of training. There is low tolerance for working through lesson after lesson to get the 60% of the information that is pertinent to the specific user's training need. Moreover, customers are sensitive to the fact that time spent training is time spent away from generating company revenue. Consequently, there is now a perceived benefit to developing adaptive CAI capabilities, as in an intelligent software capability that customizes the training interaction to specific learning needs.

Whereas a long-term objective of the research organization was to obtain support for research and development of intelligent interactive learning systems, a critical first step was to establish the organizational perception that there was some additional benefit to be gained from the more advanced technical solutions. The preliminary success of our technology transfer strategy is manifested in the observed positive impact of this project on the current use of CAI technology in the corporate training center as well as the perceived need for more advanced delivery and distribution of CAI applications.

Current Status

Despite the positive impact of the project, we have not established a firm commitment, that is, corporate funding, to continue the technology transfer of interactive learning technologies. Although there are a multitude of factors that impact an organization's commitment to research and development projects, we speculate on a couple of pragmatic acceptability issues that might explain the current level of commitment.

Credibility. The project successfully established credibility of the research organization to develop CAI applications and the CAI technology as a viable alternative to the stand-up classroom instruction. Although lack of credibility can be a barrier to obtaining commitment, strong credibility does not produce a commitment.

Feasibility. The current state of the art in instructional authoring tools is still a barrier to developing intelligent interactive learning systems. The development of authoring tools for intelligent learning systems is perceived to be a high-risk investment, as is the development of expert system development environments.

Funding. The financial structure of the organization continues to be a fundamental problem area for obtaining R&D funding for advancing the training technology capabilities. The strongest support for continued commitment to research and development lies with the training center. However, the training center budget can provide only limited support for the current CAI development projects. Most of their current CAI development funds come from new product development budgets.

Relevancy. The indirect relevance to the development of base products seems to present a significant barrier to obtaining a firm commitment from the corporate decision makers. Unless the training capability is perceived as a significant product differentiator with significant economic benefit, it is unlikely we will see significant, corporate R&D investments. Justifying claims for economic benefits from advanced training capabilities of intelligent interactive learning systems requires pointing to less tangible factors than are available for base software and hardware products.

Without addressing these identified pragmatic issues, it is unlikely that a firm commitment to more advanced research and development in intelligent interactive learning systems will be obtained.

CONCLUSION

This chapter presents a case study of successful technology transfer project that produced a significant positive impact on corporate training practices. The successful technology transfer was attributed to specific aspects of the project management and technical approach that focused on achieving acceptance from multiple users and organizational stakeholders.

The success of the project with respect to the long-term objective to obtain support for more advanced research and development of intelligent interactive learning systems has yet to be realized. Obtaining a firm organizational commitment to the long-term objective requires an R&D strategy that addresses the significant pragmatic issues associated with demonstrating credibility, proving feasibility, justifying funding, and establishing corporate relevancy. Until these issues are addressed, we do not expect to see widespread use of today's technical innovations in learning systems technology.

At the present time, the long-term objective appears to be somewhat of a quixotic goal. However, an important lesson learned from this case study is that, sometimes, smaller steps in technology transfer can have a significant positive impact when the strategic approach emphasizes acceptability of the technical solution. Furthermore, because of the current technology gap, several smaller steps may be necessary to achieve widespread use of today's advanced learning systems technologies.

REFERENCES

Dede, C. (1995, May). *Tools, situated media and fabrics: A strategy for forming alliances to develop learning technologies*. Paper presented at NIST Advanced Technology Program: Workshop on Learning Technologies, Austin TX.

Hammer, J., & Champy, J. (1993). *Reengineering the corporation: A manifesto for business revolution*. New York: Harper.

Kodali, N. (1994). *Designing computer based instruction: Iterative development and application of research literature in refinement—constraints, rationale and evolution*. Unpublished doctoral dissertation, Tufts University, Boston.

Nielsen, J. (1993). *Usability engineering*. New York: Academic Press.

Polson, M. C., & Richardson, J. J. (Eds.). (1988). *Foundations of intelligent tutoring systems*. Hillsdale, NJ: Lawrence Erlbaum Associates.

Psotka, J., Massey, L. D., & Mutter, S. A. (Eds.). (1988). *Intelligent tutoring systems lessons learned*. Hillsdale, NJ: Lawrence Erlbaum Associates.

8

AUGMENTING INTELLIGENT TUTORING SYSTEMS WITH INTELLIGENT TUTORS

E. Robert Radlinski
Michael E. Atwood
NYNEX Science & Technology, Inc.

To improve their competitiveness, industries often focus on the skill sets of their workforce. This focus recognizes that the needed skill sets are volatile and dynamic. Whereas workers at the turn of the last century might have been able to work their entire career with the skills they acquired as apprentices, current workers learn substantially new skills every several years, and the length of time that skills are viable is rapidly declining. Modernization, automation, and computerization introduce new systems into the workplace, and workers must learn to use them. These new systems, in turn, frequently redefine the tasks that can be performed so that workers learn to produce new products and services. To further complicate this problem, the skills and abilities of entry-level workers are declining significantly (Ramsey, 1988).

To address these problems, industries are spending an estimated 5 billion dollars annually just to insure that users of information technology are properly trained (Nelson, Whitener, & Philcox, 1995). As an alternative to expensive classroom-style lectures, a considerable portion of these dollars is spent on development of computer-based training systems (including intelligent tutoring systems [ITSs]), distance-learning networks, and desktop collaboration technologies. The goals of these initiatives are to deliver effective, self-paced, just-in-time instruction to the workplace.

ITS technology has been advertised as being one effective way of reinforcing instruction in procedural skills at the workplace. These systems offer individualized curricula with fine-grained feedback and analysis of student problems. They have been used effectively in a number of domains, from introductory programming to electronics diagnosis and repair (Anderson, Boyle, Corbett, & Lewis, 1990; Kurland & Tenney, 1988). They have been shown in many classes to approach the effectiveness of a one-on-one human tutor (Anderson et al., 1990). However, computer-based tutoring systems are typically limited in handling unexpected problems. Various degrees of instructor support are often required for students of all levels.

Distance-learning techniques offer a means of remote access to an instructor and an alternative way of delivering declarative instruction. Distance learning generally refers to remote links between an instructor and students. These links are often through expensive satellite downlinks offering one-way video and two-way audio and, occasionally, two-way video. Although these links are useful for delivering information and some discussion, they are not well suited for teaching procedural skills that require extensive practice. Further, studies of remote instruction suggest that the best results that such distance-learning classrooms can achieve is comparable to a typical classroom (Stone, 1992). The delivery of live instruction over satellite links typically costs about $1,000 per hour. This does not include the cost for the cameras, broadcast equipment, support staff, instructional preparation, and teacher training. In other words, many training centers are spending large amounts of money on distance-learning environments to achieve, at best, the same levels of effectiveness that occur in a typical classroom (Moore & McLaughlin, 1992).

Although frequently attempted, we see very few successful applications of either ITSs or distance-learning environments. Johnson (1988) gave what he considered the four important reasons for the lack of more successful ITSs. We contend that these reasons also apply to distance learning. His reasons were the *lack of resources, equipment, and personnel*, and the lack of *proper system developer attitudes*. We concur that these reasons easily prevent successful deployment of ITSs and distance learning in industry. However, we also note one additional reason. Both ITSs and distance-learning environments do not actively and appropriately integrate human collaboration in instruction delivery. With ITSs, the focus is typically on replacing human instructors with computerized systems. Although there is much that computerized systems can accomplish, they function most effectively when used in conjunction with human-to-human instruction. This is especially true when students arrive at a learning impasse due to a lack of prerequisite knowledge.

Distance-learning environments also do not adequately support the level of human interaction necessary to learn, especially when the individual mastery of procedural skills is required. Typically, distance-learning links tend to support a one-way monologue—from the instructor to the remote locations. However, these networks are often inadequate for student-initiated mentoring sessions.

In this chapter, we argue that, individually, both ITSs and distance-learning systems have inherent limitations, but through the proper integration of these technologies, many of these limitations can be overcome. We support these conclusions by reflecting on our experiences over the past several years in attempting to field ITSs and distance-learning systems in industry. We also report on our progress in alleviating the problems of how effectively to integrate humans and computers in the instruction delivery process and how better to promote distant collaboration.

INTRODUCTION TO THE GRACE PROJECT

The Grace Tutor (Dews, 1989), an ITS for COBOL, was developed and fielded in both academic and corporate settings. The goal of the Grace Tutor effort was to take the current state-of-the-art technology in ITSs and apply it to the advanced development of a tutor that is used in real corporate classrooms. Our focus on "advanced development" rather than "basic research" meant that what we built had to fit into a real corporate training curriculum. We felt that the technology had reached a point where ITSs were ready for "industrial strength" use and that we could migrate the existing ITS tools and adapt them to our own needs (Gray & Atwood, 1992).

The Grace Tutor was designed as a drill-and-practice tutor to teach COBOL programming as a cognitive skill. The essential, underlying assertion is that a skill must be practiced to be truly learned, not just seen in a few examples or read about in a text. Our tutoring approach required that students in training courses acquire the COBOL concepts, syntax, and the declarative knowledge through classroom lectures and then use the Grace Tutor to reinforce these concepts using the ITS exercises. After each lecture, the students worked on a set of problems that applied the newly learned concepts. The exercises also reinforced concepts learned in previous lessons in a "scaffolding" or "mastery learning" approach (Bloom, 1984).

A screen shot of Grace as it was used in the classes is shown in Fig. 8.1. The large window, labeled Cobol Code, is where students generate their solutions. The completed syntax is represented by uppercase text, and the place holders and subgoals of the syntax to be completed are represented in lowercase, italicized text. The current subgoal is enclosed by a rectangular cursor. The Menu window to the right of the code window contains context-sensitive menus of COBOL phrase templates; the Interactor window at the bottom displays the feedback generated by the tutor. A Dialogue window appears when a name or a value is required as typed input by the student.

At the beginning of each student-tutor interaction cycle, the computer highlights one of the unfinished subgoals. This is shown in the Cobol Code

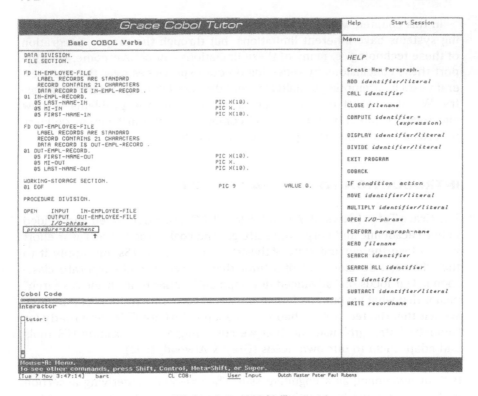

FIG. 8.1. Grace COBOL Tutor.

window in Fig. 8.1 where the *procedure-statement* has been selected. Although students had the option of selecting any unfinished subgoal to work on, a large majority preferred to work in the default (i.e., top-down, left-to-right order). In this example, the student is required to pick the next correct COBOL statement from the list of statements given in the Menu window. This cycle generally alternates between picking a verb from the menu and entering its associated arguments.

If the action is correct, the code appears on the screen. If the action is incorrect, error feedback is presented in the lower Interactor window. Therefore, as long as the student is progressing correctly (and there is discussion following of just how the Grace Tutor decides whether the student is right or not), the tutor stays in the background and acts as a structured editor. As soon as an incorrect choice is made, the Grace Tutor intervenes by blocking the student and presenting feedback.

To evaluate student responses, Grace employs an ideal student model. The *ideal student model* in the Grace Tutor is a cognitive model or simulation of a good student solving a COBOL exercise. It simulates an expert system in that it generates solutions to COBOL exercises. It differs, however, in that it generates solutions that a good beginning or intermediate student does. These are not necessarily "optimal" solutions as there may be several ways to solve an exercise. Therefore, in the purest sense, it is not an expert model but, rather, a model of a good student.

The Grace Tutor is capable of generating any "reasonable" solutions expected from students. However, the ideal student model is static in nature. In the rare case where a student demonstrated a unique but reasonable solution, adjustments to the model were required.

The Grace Tutor's student model is a production system implemented in PUPS, the Penultimate Production System (Anderson & Thompson, 1989), in which each production corresponds to a knowledge structure or "chunk" that the student has learned. At each step in the problem, the production system finds all possible next steps that lead to a correct solution. It then checks the next input by the student to see if it matches any of the possibilities. It progresses along the solution paths, correcting the student when a match cannot be found, until a solution is complete.

As the productions are traced through the student model, another model is built on the current student's knowledge. If a production is found that corresponds to the current action of the student, the probability that the student has acquired the corresponding piece of knowledge is incremented. If a production cannot be found, the tutor selects the production that it would have used in the ideal solution and decrements the judged probability that the knowledge had been learned. This body of weighted productions is a trace of the student's knowledge at any moment, and the process by which these weights are attained is what we call *knowledge tracing*.

The lessons in the Grace Tutor are organized into topics such as "variable declarations" or "tables" that roughly correspond to classroom lectures or sections in a standard COBOL text. At this level of modeling, the lessons are distinguished by the new productions that occur. Knowledge tracing is only applied to the new productions, because the student is not allowed to progress to a new lesson until the tutor judges that all productions have been thoroughly learned.

Together, knowledge tracing and the ideal student model provided the Grace Tutor with a very fine-grained and detailed assessment of students.

GRACE ON TRIAL

The Grace Tutor was built keeping in mind the cognitive principles that we believed were important for teaching and modeling. We built an interface that we felt supported the student and designed a curriculum that built up the skills appropriately, according to the best principles we could find. However, demonstrating whether the Grace Tutor was actually useful and effective could only be done in a real classroom. This section presents the results of three trials: one in a college programming course and two in industrial courses for programmers learning COBOL as a second (or nth) language.

The SUNY Trial

The Grace Tutor was first tested during a course at the State University of New York (SUNY) at Purchase. Eight students were selected by the instructor to use the Grace Tutor at the halfway point in the course because their low scores on the midterm exam and laboratory exercises put them in danger of failing the class. The Grace Tutor students met one night a week for approximately 3 hours over 3 weeks, while the other students participated in the normal lab sessions. We worked with the instructor to modify the tutor's curriculum to match that of the class.

Comparing their midterm grades to their final exam grades, the Grace Tutor students improved from a 63% average grade to a 76% average grade. We were encouraged with these results because the Grace students were able to keep pace with the rest of the class even though they started the second half of the course with a substantial deficit. Of the four Grace students who had failed the midterm exam, all were able to pass the course, two with Bs and two with Cs.

Although the improvement in scores was promising, the long-term value of working with the Grace Tutor for these students was an increase in confidence and motivation. When they first arrived, these students were convinced that they were going to fail the course, and none felt confident in their programming skills. After using the tutor, they were excited about the amount they had learned and were confident that they were prepared for the final exam. Some of the comments collected on questionnaires were: "Working with the tutor helped me understand much better overall what the instructor was trying to talk about in class," "Grace Tutor has improved my knowledge and ability to fully comprehend COBOL programming *100%*," and "I reread the class notes and chapters assigned to us and everything just 'clicked'."

The results of the SUNY trial were encouraging and gave us evidence that the Grace Tutor is useful for helping beginning students, particularly those needing remediation. However, within industry, many of the students that are likely to use the Grace Tutor are ones who are not complete novices at programming and may come from much more diverse backgrounds than college students. We felt that it was important to test this group of target users and so we conducted two trials in industrial settings with more experienced programmers.

The Met Life Trial

The first trial in an industrial setting was conducted at Metropolitan Life Insurance Company in New York City. This was a 2-week course entitled "COBOL as a Second Language." All students were required to have programming experience, though we found that the actual experience varied greatly. The class was held every day and, generally, there was a lecture in the morning and programming exercises in the afternoon.

Twelve students in the class were divided into two groups of matched pairs. The matching was done on several measures: a standardized Programmers' Aptitude Test score (SRA, 1986), a COBOL knowledge test, and a questionnaire that asked about other programming languages and background. Even though the students were supposed to have programming experience, several had only used packages such as spreadsheets or had learned an in-house language called Metropolitan English Language or MEL. A few had never used a mouse or a windowed interface. Thus, we had a fairly wide range of ability and experience in the group. Pairs were made and then a coin toss decided which group would use the Grace Tutor.

Both groups attended lectures together. The Grace Tutor group received a brief (about 20 minutes) orientation to the interface and then worked on the tutor exercises, while the other group worked on nine "case problems" that were COBOL exercises done on the regular mainframe system. The Grace group also worked on three of the case problems. This decision was made by the instructor primarily so that those students unfamiliar with the VMS operating system and compiler had adequate experience by the end of the class.

The posttest was the same COBOL test that was given as a pretest. This consisted of five programming problems that increased in difficulty. It was a "speed test" in that the students were given one hour to work on the test, and none of the students finished all the problems. Although there were too few participants to do a statistical analysis, results indicate that the tutored participants did somewhat better than the nontutored group with the Grace group mean being 62.2% (*SD* 22.9) and the non-tutored group scoring 51.9% (*SD* 27.7).

These test results were encouraging and indicated that the Grace Tutor was at least as effective as the lecture plus nine case problems and took less time. Also, the instructor had to spend less time with the tutored students. This was predictable because the tutor provided feedback to the tutored students, but the instructor was the only source of feedback for the nontutored group. The comments from the students indicated that, overall, they liked the experience of working with the Grace Tutor. We did note that the more experienced the programmers, the less they liked the tutor. We have more to say about this observation later in this chapter.

The New York Telephone Trial

However, we still had not conducted a trial with the primary target group for which Grace was developed. New York Telephone (NYT) trained several hundred COBOL programmers each year. Because NYT was a sponsor of the NYNEX Grace project, we wanted to be sure that Grace was useful for the students within NYT. Therefore, we scheduled a trial at the Computer Technology Training Center.

NYT had not taught a course in COBOL as a second language, but they were very interested in offering one. Because they did not have a curriculum or instructor already at hand, the Met Life instructor offered to serve as the instructor for them. We felt that this was very helpful for our evaluation as well because we could attempt to duplicate as closely as possible the previous Met Life trial. The course was taught in exactly the same manner except that the 10 days spanned 3 weeks because of holidays. Also, there were a few very minor differences between the Met Life standards and the NYT standards, but these did not effect the Grace Tutor curriculum.

There were eight students in the NYT course. Again, the students were matched as much as possible on the same pretest criteria used in the Met Life trial. Because of the small number of students, the pairs were more difficult to make. Therefore, we had an instructor from another class who was familiar with all the students help in the pairing to make what she felt were groups of relatively equal ability. Again, the tutor/nontutor decision was made by tossing a coin. In general, the NYT students were more homogenous, basically being a more experienced group than the Met Life group. This is reflected in their less-varied scores on the SRA Programmer Aptitude Test. All of them were programming at NYT in their current jobs, mostly in PL/1. None had any training in COBOL.

After the class, we gave the same COBOL posttest that consisted of five programming problems. In addition, the NYT staff administered the Berger COBOL Test, which is a standardized test of COBOL knowledge. It is multiple choice and mostly covers syntax with a few logic questions. Again, the Grace Tutor group came out with a slight overall advantage, though the variance was great and the numbers small. On the Berger test, the Grace Tutor students' average score was 63.8 (*SD* 23.9) and the non-tutor score was 53.8 (*SD* 30.7). On the COBOL test, the Grace Tutor students' score was 70.0 (*SD* 12.4) and the non-tutor score was 68.2 (*SD* 7.7). The COBOL programming posttest solutions were graded blind by the instructor (i.e., students' tests were identified only by a number).

Again, the tutored students felt that the Grace Tutor had helped them learn. However, these students were a more experienced group of programmers than before and had much to say about what they did not like as well.

We learned a great deal from their comments and our observations. These observations are discussed in the Lessons Learned section.

OBSERVATIONS

Although we did not achieve a final deployment of Grace, the trials gave us the opportunity to observe how the students used the system. In this section, we describe two observations from the Grace developments. These observations—the need for human intervention and the need to tutor in the context of use—combined with a parallel effort to support distant collaboration led to the design of the second generation of Grace, which we call DIME. We introduce the concept of DIME through a discussion of these observations.

Intelligent Tutoring Systems With Human Intervention

ITSs such as Grace have generally been used to reinforce classroom or textual knowledge. One problem with these tutors is that when a student encounters difficulties on a problem or is lacking a piece of prerequisite information, the tutor cannot offer the flexible help normally available through an instructor. For example, we encountered a student who did not understand negative numbers, a concept that was embedded throughout a number of Grace problems. As with many other concepts, we had assumed that all students brought a certain set of prerequisite knowledge to the course, including the understanding of integers. The Grace Tutor was not capable of remedial instruction on this concept. Given this impasse, the student had to rely on the COBOL instructor for supplemental help.

In fact, it was common to see students raise their hands while using Grace. Students frequently asked the instructor to clarify a problem, to verify a solution, or to ask for remedial instruction. Intervention was also required when unexpected problems arose, such as the negative number example. Although Grace did reduce the need for instructor intervention, it came nowhere close to eliminating the need for this collaboration. Remedial instruction still required human intervention.

We also observed frequent collaboration among the students themselves. In each of the trials, students quickly established informal networks among themselves to exchange ideas and questions. They frequently walked from workstation to workstation to observe, discuss, and question each other. These discussions involved a range of issues—from the check of another's progress to minimentoring sessions. Both instructor–student and student–

student discussions proved to be an integral part of the learning process with our students.

We began to consider how future versions of Grace could better support these social interactions. These interactions are an important activity in the learning process and are especially important when students are in search of help.

Tutoring in the Context of the Problem

To our surprise, the Grace students quickly developed a considerable intolerance with our help system. When a mistake was made, students were always prompted by a help message. Three levels of help were available for each step in a solution. On the first mistake, a simple reminder of the task was stated. A more informative description of the current concept was given if the student erred a second time. If the student missed the concept on a third try, Grace always provided the correct answer.

We observed that students soon became annoyed with Grace's explanations. The better students did not want to be given any hints or answers. They simply wanted to be notified that a mistake was made. They preferred to solve the problem on their own without any assistance. Secondly, almost all students had difficulty in understanding any of the explanations. Although the Grace Tutor always gave the next correct part of the solution, it typically gave the explanation out of the students' context. Still confused, students again solicited help from the instructor or from other students. We observed that both instructors and students are excellent diagnosticians and are best capable of providing meaningful explanations.

Explanations are most meaningful when given in the context of each student's current solution. However, Grace only allowed the instructors to see the current state of the solution when summoned for questions. When asked for help, instructors usually spent several moments reviewing or probing into *how* the students arrived at their current state of the solution. Frequently, students were asked to reconstruct their solution, either verbally or on paper. Often, the students misrepresented this reconstruction and frequently forgot errors and assumptions that they had made along the way. Finally, and perhaps most importantly, students summoned help only after they had made multiple mistakes.

We noted that a facility that provides the instructor with a fuller analysis of the problem solution is more beneficial than a simple snapshot of the solution's current state. This type of problem replay facility could be a logical extension of our Grace Tutor because student data was constantly being collected for our knowledge-tracing model. By reconstructing a replay of any solution, the instructor could see *how* the current state of the solution was developed and could ask questions as to *why* the student made various choices. Seeing the student's errors also provides considerable insight into a student's level of understanding. By seeing how a student arrives at a certain point in a solution, an instructor may better understand *where* a student is having difficulties.

Knowledge Assessment

The Grace Tutor included a knowledge-tracing tool to instantaneously view the current status of each production. This status included data about how many times a production (concept) was tried, the number of successful applications, and the current probability that the production was learned. However, this facility was difficult for instructors to understand as the granularity of measure and perspective was not the same as the instructor's. In the context of programming, instructors often look for a breakdown in logic, not in the misapplication of a syntax rule. In this situation, the students' knowledge was being traced from two different perspectives, the instructors' and the tutoring system's. We observed that, although both are important, the instructors' perspective prevailed in such discussions.

Secondly, the Grace Tutor's knowledge-assessment mechanisms were not well understood nor accessible to the instructors. This situation often resulted in considerable confusion, especially when students asked for Grace's graduation criteria for any given lesson. This became more problematic when students felt that they had demonstrated competency in the current lesson but were still required to do remedial problems. A lack of a hierarchical structure among the productions often allowed "minor" productions to overinfluence the knowledge-assessment process. Because the instructors did not have an "override" option, students were required to do remediation, sometimes when it was unnecessary. To compound matters, the students themselves were prohibited from viewing their current status, which resulted in further frustration.

Thirdly, the Grace knowledge-assessment facility was perceived as too rigid. Students frequently made typographical errors while entering problem-specific variables or when choosing syntax items from menus. These typographical slips resulted in considerable frustration as the students did not have the opportunity to review and correct their input *before* the Grace Tutor began its assessment. As mentioned earlier, each error invoked the tutor's feedback mechanisms, and various degrees of *unsolicited* help were generated. Students soon discovered that each of these slips typically resulted in having to do more exercises. Moreover, we concluded that many of these slips were caused in part by the Grace Tutor's somewhat cumbersome interface (McKendree, Radlinski, & Atwood, 1992).

Collectively, the aforementioned lessons provided us with additional usability and pedagogical requirements that will improve the effectiveness of the second generation Grace Tutor. The following section describes how we have begun to implement these requirements in a new system architecture called DIME.

Technology Transfer

As we mentioned earlier, one of the strategies of the Grace project was to adopt state-of-the-art tutoring technology developed in academia, modify it to our needs, and integrate it into our corporate classrooms. By not having to design and develop a new tutoring architecture, we saved considerable time in its development. We also viewed the delivery of the next generation of computing systems, hardware and software, as having a positive influence on upgrading the training organization's computing environment. What we failed to consider, however, is the negative side effects of this decision.

When we adopted the PUPS production system, we also adopted its basic hardware and software components. Although we did migrate from a Symbolics to a McIvory platform, we still continued to develop the Grace Tutor in LISP, to use a multiple windowed interface, and to rely on the mouse as part of the input process. However, this style of interface and the concept of stand-alone processors had a significant impact both on the students and the corporate training organization.

As was the case in most training organizations, the systems that students used were a bit antiquated, especially in comparison to the Grace platform. Their terminals consisted of TTY interfaces, typically had single-window interfaces, and were all supported and linked by mainframe computers. Many of our students had not used a mouse before, a discovery that forced us to hold mouse orientation sessions at the beginning of each of the trials. Students also questioned how a PC could be powerful enough to understand how to tutor without the use of a mainframe. They were also disappointed that they could not compile, execute, and debug their Grace solutions on their mainframe systems.

The reaction from the training organization was even more negative. From the business sense, the training center was completely surprised by our introduction of "Mac PCs" into their "IBM" classrooms. Two questions were typically asked when we introduced our systems—"How much?" and "Is this system compatible with our IBM mainframes?" After giving the appropriate answers ("$30K" and "No," respectively), all further discussions of the future porting of Grace to IBM systems fell on deaf ears.

What we unexpectedly discovered is that we had introduced a system generation gap in the classrooms, a gap that was met with significant resistance. Although the students quickly adapted to Grace, the training staff did not. In part, the instructors were intimidated by the "teaching power" of Grace and perceived the tutor more as a competitor than a teaching aid.

The concept of computer-based rather than instructor-led tutoring did *not* fall on deaf ears. Although cost and system incompatibility issues were the officially cited reasons for the lack of deployment, we believe that our omission of training the staff contributed to the demise of the Grace Tutor.

The results of our Grace trials taught or reaffirmed four basic principles. First, it was possible to teach the procedural knowledge of programming using ITSs such as Grace. Second, humans are still the best providers of declarative and causal knowledge. They are especially good at diagnosing learning impasses and at providing meaningful explanations in the context of the students' needs. Third, students want and need to interact with others during the instructional process. Learning is a social process and the introduction of technology into the learning environment must promote these social activities. Finally, the introduction of new instructional tools into classrooms must be gradual and planned. In addition to the pedagogical concerns, business and cultural issues need to be considered, as does sufficient training.

In the following sections, we demonstrate the adaptation of these principles in our continuing work in training technologies.

DIME: DISTRIBUTED INTELLIGENT MULTIMEDIA EDUCATION

The Concept

These principles radically changed our tutoring systems research. Rather than continue to explore ways of incorporating more declarative and causal information into Grace, we began to focus on providing this knowledge using the richest source available—the human. In this section, we describe the conceptual architecture of a new initiative that we call DIME (Distributed Intelligent Multimedia Environment).

The DIME concept represents an attempt to integrate three cognitive systems: ITSs, instructors, and other students. This triad is represented in Fig. 8.2. The apex of this system remains an ITS, such as Grace. The goal of the tutor is to continue providing individual guidance and judgment to students who are applying newly learned concepts to real problems. This style of mentoring during drill and practice exercises is difficult to achieve with human instructors in *any* classroom. We continue to see the need for such systems if learning is to be truly effective.

The two other points of our cognitive triad represent the integration of two sources of human knowledge: instructors and other students. Instructors, as indicated throughout our trials, are best at providing declarative knowledge, diagnosing student problems, assessing learning, and providing

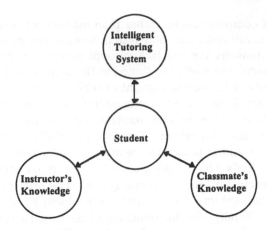

FIG. 8.2. Dime's cognitive triad.

information in the context of students' needs. Similarly, student resources satisfy the social interactions that are necessary among students. Such interactions promote teamwork on larger projects, offer the opportunity for peer mentoring, and often foster competition.

The goal of the DIME initiative is not only to provide these three sources of knowledge in a desktop learning environment but also to insure that both student and domain knowledge is appropriately shared *among* these resources. We describe our proof of concept implementation of this framework and further describe our vision in the following sections.

DIME

We begin our discussion of DIME with a description of the revised Grace Tutor. Although only a portion of the COBOL curriculum has been ported to the DIME system, we are convinced that the new system supports the necessary functionality for corporate training.

The tutor has been transformed in many ways. It currently runs on a PC platform, under both Windows and OS/2, platforms that are affordable and that integrate nicely into the corporate training environment. Although we continued to use a production system paradigm for the necessary modeling, commercially available software tools have been used for implementation. The tutor's interface has also been radically changed to more closely reflect the needs of programming (McKendree et al., 1990). The interaction is reduced to a single window, as shown in Fig. 8.3. We believe that the new implementation of the tutoring component corresponds more closely to the actual environment in which the programmers work.

The most significant enhancements in DIME are reflected in the menu structure. The Instructors and Students menu items represent the addition

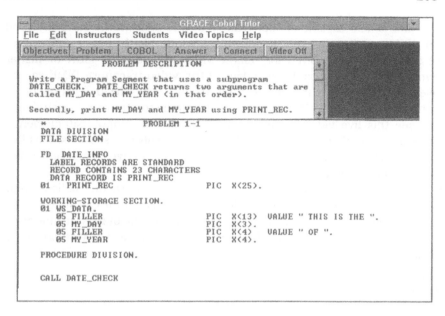

FIG. 8.3.   New DIME interface.

of a computer-supported cooperative work (CSCW) facility that is designed to promote both instructor and student collaboration from any location. DIME's CSCW facility, which we call Interactive Coach, supports human collaboration over a variety of communication medians. The Interactive Coach is an electronic version of a mentor looking over a student's shoulder and providing support when asked. When human intervention is required during the course of an exercise, a student can simply choose a phone number from a menu of instructors or other students. A phone call is initiated by the Interactive Coach, and either an audio or a video connection is made, depending on the capabilities of the student's PC. Once a connection is made, the Coach provides both parties with the ability to discuss the student's problem(s) in the context of the ITS problem. This facility is described further later.

At a minimum, the Interactive Coach currently requires the addition of a digital modem to the PC. All collaboration activities are provided over standard telephone lines (LAN/WAN and ISDN implementations of the Interactive Coach are planned).

The hypertext system of the original Grace Tutor has been augmented with a library of multimedia presentations. The Video Topics menu permits students to preview or review lectures and presentations while solving the tutor's exercises. The viewing of specific lessons can be initiated by the student, the ITS, or by a remote instructor when the Interactive Coach has been activated.

DIME is also capable of supporting multiple bandwidths currently used for desktop video conferencing. The blank window in the upper right corner of the screen is reserved for remote video provided by ISDN or higher bandwidth connections. Independent of the video transport, DIME can be used to receive live training lectures from a centralized location or to support one-on-one video discussions with an instructor or other students. We see this live video capability as necessary to support supplemental drawings and other visual aids during remote mentoring sessions. This live video capability can be used along with the Interactive Coach and other multimedia materials.

Collectively, these components allow students to practice skills at their own pace and schedules with the aid of the ITS. The tutor can suggest that they review lectures that can be stored either on videotapes on-site or as a video library available through the workstation. We also give students the means to contact an instructor or other students on a per-need basis.

DIME as an Instructional Tool

We envision using DIME in the geographically distributed training areas at each NYNEX location that provides video capabilities, much like a traditional teleconferencing room. This allows lectures to be delivered in a timely fashion to employees. Employees could then return to their individual workstations or remain in the training area to work on problems with the ITS. Each workstation is equipped with the tutor and the ability to connect with a human expert through an audio and data line. In other words, our system does not eliminate the need for instructor-led lectures in some centralized fashion but concentrates on improving the training that follows the introduction of declarative knowledge.

After having attended a lecture, the students begin work with the conventional ITS. They work on the problems presented, receiving feedback and an individualized curriculum. They can review information about COBOL through the online hypertext documentation that can include material from an online video library as well as audio, textual, and graphic information. As long as students feel that they are progressing and understanding the material, the session will proceed as in any ITS session.

When students run up against a problem that requires human intervention, they simply click on a list of phone numbers that can be dynamically updated to show who is available at the time. A phone call is initiated, the instructor or another student answers, and either an audio connection is made, or if the workstation allows, a live, face-to-face conversation is initiated.

Although this conversation is useful, it is also necessary for the remote individual to be able to see the work that the student has done so far. This ability to have a "shared problem space" is accomplished by sending data about the problem being worked on through the data link to the instructor. The student can click on another button that sends the history of the current solution to the remote workstation. This problem transfer allows the individual at the remote workstation to watch an "instant replay" of the solution and to see exactly what the student has done, including all intermediate steps. The replay can be paused at any time to discuss the logic, syntax, or alternatives of any step and then be continued to see the remainder of the solution.

At the conclusion of the transfer of a student's solution, the two PCs are "synchronized" that is, a shared workspace is created to allow *both* the student and the remote party to continue to *interactively* complete the solution. The two can continue to discuss alternatives or mistakes, backup, restart the problem, finish the current solution, or even start a new problem to emphasize a particular concept. Until the Interactive Coach link is broken, total interactivity is sustained—both can use their own mouse or keyboard, view any hypermedia support, and discuss the problem until both feel that the student is again able to progress independently. Once the student feels comfortable working alone again, the Interactive Coach link is terminated, and the student continues working under the control of the tutoring system, as before.

Future plans include an asynchronous version of this scenario. If the remote party is unavailable, DIME will allow the student to send the problem history along with a voice annotation to the remote party where it can be reviewed as an e-mail/voice message at a later time. In similar fashion, an asynchronous response from the remote party will be supported. In this scenario, problems can be discussed asynchronously yet in the context of the student's needs.

Benefits of DIME

In general, we think that the most valuable enhancement to our tutor is the shared workspace component (our Interactive Coach). This facility enhances computer-based tutoring by providing human tutoring inexpensively. Students, especially students in distance learning-environments, have more flexibility to work on the training material *where* and when they want and still have the opportunity to access an instructor *when* needed. Students will not have to travel to a remote location for, perhaps, several days. Yet, they still have access to an instructor and can share the work as if the instructor were there "looking over their shoulder." This characteristic of being able to choose *when* to receive instruction has been shown in distance-learning studies to be the major factor in determining student success (Stone, 1992).

Student errors provide considerable insight into students' level of understanding. Our earlier version of Grace allowed instructors to see only the current state of the solution when summoned for questions. Again, the Interactive Coach component of DIME provides a full replay, including errors and restarts, of any problem. By seeing how a student arrives at a certain point in a solution, an instructor may better understand *where* a student is having difficulties. Often, a student will summon help only after multiple mistakes have been made. The Interactive Coach allows the instructor to conduct a full analysis of the problem by providing more than just a snapshot of the solution's current state.

Knowledge tracing can be supported by extending the Interactive Coach component in a noninteractive way. By archiving the student's solutions in a local server, we can allow the instructor to spot check any student at any convenient time. By replaying the student's problems from a repository, the instructor will be more capable of detecting not only what productions have been learned but also *how* they were applied. This feature also begins to address one of our previous problems—that of teaching optimal and/or alternative solutions. In programming as well as in many other procedures, students will learn one technique and then stay with it. This enhanced knowledge-tracing capability will better demonstrate and allow enforcement of students' learning alternative solution paths.

In all of our previous tutor trials, students complained about the restrictiveness of our knowledge-tracing mechanisms. Sometimes, a single violation of a single production resulted in students doing one or more remedial problems. During each of our trials, we spent considerable time making adjustments to our production status data and sometimes relaxing the graduating criteria by brute force. We see the problem-replay facility as a useful tool to fine tuning the tutor's models. We speculate that with a constant use of the replay facility, *instructors* can better understand the implications of any individual production and can fine tune our system's success criteria. Secondly, by passively reviewing archived solutions, instructors can better detect both individual and classwide trends for misunderstood concepts and other problems. Trend analysis is a difficult task when using only the production statistical data of individual student models.

The inclusion of video technology in tutoring systems focuses on a higher level of pedagogical benefits. The affordability of video cards and CD-ROM technology provides the capability of viewing prerecorded minilectures. If the student becomes stuck or shows a weakness with respect to a certain set of productions, the capability of the tutor or an instructor to recommend the review of previous lectures may assist in remediation. This facility is also important when a student misses a lecture because of work priorities, sickness, and so on.

Originally, we envisioned the use of live video to be reserved for unanticipated situations. As in the case where the student did not understand negative numbers, instructors can focus the camera at the chalkboard, conduct a review or lecture, and solve example problems in a one-on-one capacity until the student's understanding is at an acceptable level. We see this level of human intervention as unnecessary for all mentoring sessions, but it is still required for the more difficult mentoring sessions.

However, desktop video conferencing technology has become more affordable, as has the cost for communications such as ISDN. Therefore, we continue to incorporate more of this capability in our DIME prototype. Even though video may be useful for showing external objects such as diagrams, notes, and so forth, we believe that allowing the instructor and student to see the same workstation screen is *the* critical characteristic for electronic mentoring.

SUMMARY

The fundamental principle of DIME is to support learning by doing, not learning in isolation. As our trials demonstrated, computer-guided problem solving often requires human intervention as a means of supplementing the limitations of ITSs. The goal of DIME is to establish a partnership of multiple cognitive sources—ITS, instructors, and peers—that can be used during problem-solving tasks.

We feel that DIME is a promising alternative for corporate training. This hybrid system approaches the goals of "apprenticeship" and "situated learning" without requiring a human instructor to be available at all times for one-on-one tutoring. The ITS can serve as a "mentor" for routine practice, practice that may be quite complex. When a nonroutine situation arises or the student becomes confused or feels the need to discuss any related issue, human assistance is made available in the form of instructors, subject-matter experts, or other students. This assistance can be available from the workplace or after normal work hours in either synchronous or asynchronous form. DIME not only supports human interactions when the student needs them but also focuses these collaborations in the context of the student's world.

We also believe that DIME's ability to support human collaboration during the learning and application of new concepts is a logical extension to ITSs. However, its success is heavily dependent on its acceptance by corporate training organizations. As in the case of the Grace Tutor, systems such as DIME will have a significant impact on instructors and the corporate training culture at large. First, instructors will need to learn how to be effective lecturers in electronic classrooms. Second, instructors will need to assume more of a role as educational diagnosticians and electronic facilitators. Third, student assessment will require the analysis of a replay of the entire problem (errors and all) and not just a judgment of the final answer. Fourth, questions will no longer be limited to scheduled classroom sessions but will be posted throughout the day (and night). Finally, a deeper knowledge of the subject matter will be required of instructors, sometimes with the assistance of system experts.

As in the case of Grace, unless tools such as DIME are properly integrated into the corporate training organizations, they will be perceived as technical novelties rather than educational tools. Training, training, and more training will be required for all levels of the corporate training staff. New business cases and training strategies, based on quantifiable needs and benefits analyses, are required to justify the costs of ITSs and DIME-type enhancements. And, as in any other advanced development, risks not only need to be identified and controlled, but also tolerated.

To a large degree, the introduction of new technologies such as Grace and DIME into the corporate mainstream is complicated by the necessitated marriage of two extremely different cultures. Corporate training organizations are typically conservative, whereas research organizations are not. Success metrics differ vastly between the two, as do educational philosophies, budgets, and resources.

We conclude with no prescriptive solutions that resolve these corporate and cultural barriers. We simply caution those who attempt to span them. When we look at the potential of the DIME, we are left with this puzzle: How can such an architecture *not* have a profound impact in the corporate and distance-learning classrooms? We again review Johnson's (1988) four reasons for success and append our final criterion: *the lack of effective transformation of the organization and people involved.*

ACKNOWLEDGMENTS

The original Grace development team also included Wayne Gray, Bart Burns, Anders Morch, Thea Turner, and Haresh Sabnani. We thank New York Telephone Co., Metropolitan Life Insurance Co., and SUNY-Purchase for providing students for these trials. Thanks to John Anderson and his lab for providing software and much assistance. Special thanks to Barbara Jaslow for her time and effort spent teaching these classes and contributing to the design of the Grace curriculum.

Jean McKendree was a codesigner of the DIME system. Her work in bringing the DIME concepts to fruition was invaluable. We also wish to recognize Jan Stein and Connie Carlson for their work on the Interactive Coach component and Michael Villano for his contributions to improving the DIME interface.

REFERENCES

Anderson, J. R., Boyle, C. F., Corbett, A. T., & Lewis, M. W. (1990). Cognitive modeling and intelligent tutoring, *Artificial Intelligence, 42*, 7–49.

Anderson, J. R., & Thompson, R. (1989). Use of analogy in a production system architecture. In S. Vosniadou & A. Ortony (Eds.), *Similarity and analogy reasoning* (pp. 267–297). New York: Cambridge University Press.

Bloom, B. S. (1984). The 2-sigma problem: The search for methods of group instruction as effective as one-to-one tutoring. *Educational Researcher, 13*, 3–16.

Dews, S. (1989). Developing an ITS in a corporate setting. *Proceedings of the Human Factors Society 33rd Anual Meeting* (pp. 1339–1342). Denver, CO.

Gray, W. D., & Atwood, M. E. (1992). Transfer, adaptation, and the use of intelligent tutoring technology: The case of Grace. In M. Farr & J. Psotka (Eds.), *Intelligent instruction by computer: Theory and practice* (pp. 179–203). New York: Taylor & Francis.

Johnson, W. B. (1988, November). *Intelligent tutoring systems: If they are such good ideas, why aren't there more of them?* Paper presented at the 10th Interservice/Industry Training Systems Conference (I/ITSC), Orlando, FL.

Kurland, L. C., & Tenney, Y. J. (1988). Issues in developing an intelligent tutor for a real-world domain: Training in radar mechanics. In J. Psotka, L. D. Massey, & S. A. Mutter (Eds.), *Intelligent tutoring systems: Lessons learned* (pp. 119–179). Hillsdale, NJ: Lawrence Erlbaum Associates.

McKendree, J. (1990). Effective feedback content for tutoring complex skills. *Human-Computer Interaction, 5*(4), 381–413.

McKendree, J., Radlinski, B., & Atwood, M. E. (1992). The Grace tutor: A qualified success. Proceedings of *Intelligent Tutoring Systems '92*. Published as *Lecture Notes in Computer Science*. In C. Frasson, G. Gauthier, & G. I. McCalla (Eds.). (Vol. *608*, pp. 676–684). New York: Springer-Verlag.

Moore, C. E., & McLaughlin, J. M. (1992, February). Interactive two-way television: Extending the classroom through technology. *T. H. E. Journal*, pp. 74–76.

Nelson, R. R., Whitener, E. M., & Philcox, H. H. (1995, July). The assessment of end user training needs. *Communications of ACM*, pp. 27–39.

Ramsey, N. (1992). Saving the schools: How business can help. *FORTUNE*, November 16, pp. 147–174.

Stone, H. (1992, April). *"Multimedia that 'Works': Beyond random access"*. Paper presented at International Distance Learning Conference (IDLC), Washington, DC.

CASE STUDIES
FROM GOVERNMENT

PART

III

"A PROPHET WITHOUT HONOR ..." CASE HISTORIES OF ITS TECHNOLOGY AT NASA/JOHNSON SPACE CENTER

R. Bowen Loftin
University of Houston

As the biblical quotation suggests, developers of new technologies are often least welcome and least appreciated by the very organizations in which they are found. Every mature organization of size has a culture that is inertial, that is, most organizations have innate defense mechanisms that resist change. Without arguing whether this is good or bad for the organization, this chapter is intended to relate case histories of efforts to place intelligent tutoring systems (ITSs) into the operational environment of the National Aeronautics and Space Administration (NASA) and to draw conclusions, based on these cases, that can aid ITS developers in increasing the probability that their systems will be accepted and used by the target organization.

Two projects are described in detail, contrasted, and used as a basis for establishing a "recipe" for successful development and user acceptance of ITSs. Brief attention is also given to the process of transferring technology from the government to the private sector, with special emphasis on the nontechnical issues inherent in such transfers.

THE CONTEXT: NASA'S TRAINING ENVIRONMENT

Historically, NASA's training approach has focused primarily on simulation-based training for both crew and ground-based personnel. This usually takes the form of immersion in water with physical replicas of flight hardware (in the Weightless Environment Training Facility or the Neutral Buoyancy Laboratory) and the use of physical simulators (Single-System Trainers and the Shuttle Mission Simulator) for astronauts. Flight controllers are trained at their consoles in the Mission Control Center (MCC) during integrated simulations and, in the case of novices, by observing experienced flight controllers during actual missions. Given the infrequent number of Space Shuttle missions (seven to eight per year) and the high cost of integrated simulations (especially those that are not flight-specific), training of novice flight controllers can take years.

Another important feature of NASA's training environment is its large focus on astronaut training. Training of ground support personnel, although a part of the charter of the internal training organization, is clearly of lower priority than astronaut training, receives a disproportionately smaller share of available resources, and is often left in the hands of the flight controllers themselves.

THE FIRST NASA INTELLIGENT TUTORING SYSTEM

In the Beginning . . .

The first attempt at developing an ITS for use in NASA began at the Johnson Space Center in 1986. At that time, several positive factors seemed to be in place: Space Shuttle flights had been suspended in the wake of the Challenger accident, freeing operational personnel from the immediate pressure of supporting the flights; the U.S. Air Force program to prepare its own flight operations personnel was reaching a conclusion, with a concomitant transfer of experienced personnel and impending loss of hard-won corporate knowledge; the process of moving more flight operations responsibility from civil servants (a stable workforce) to contractors (a more dynamic workforce) was accelerating, resulting in an increasing need to train more new personnel faster than ever before; and there appeared to be a sincere interest, on the part of middle management, in exploring the potential benefits of new technology. Thus, conditions, on the surface, seemed favorable for developing and deploying the first NASA ITS.

The Training Context

The Mission Operations Directorate (MOD) at the NASA/Johnson Space Center is responsible for the ground control of all Space Shuttle operations. Those operations that involve alterations in the shuttle's orbital characteristics are guided by a Flight Dynamics Officer (FDO) who sits at a console in the "front room" of the MCC. Traditionally, the training of the FDOs (called "fidos") in flight operations has been carried out through the study of flight rules, training manuals, and, principally, on-the-job training (OJT) in integrated simulations. Using this approach, from 2 to 4 years were normally required for a trainee FDO to be certified for many of the tasks for which he or she was responsible during Space Shuttle missions. OJT is highly labor intensive and presupposes the availability of experienced personnel with both the time and ability to train novices. In 1986, following the Challenger disaster, the number of experienced FDOs was being reduced through retirement, transfer (especially of Air Force personnel), and promotion. As a supplement to the existing modes of training, the Orbit Design Section of the MOD requested the Artificial Intelligence Section (AIS) of the Mission Support Directorate to assist in developing an autonomous ITS (dubbed an intelligent computer-aided training or ICAT system). After extensive consultation with ODS personnel, a particular task was chosen to serve as a proof of concept: the deployment of a Payload-Assist Module (PAM) satellite from the Space Shuttle. The task was complex, mission-critical, and required skills used by the experienced FDO in performing many of the other operations that were his or her responsibility.

Objectives

The training system was designed to aid novice FDOs in acquiring the experience necessary to carry out a PAM deploy in an integrated simulation (see Fig. 9.1). It was intended to permit extensive practice with both nominal deploy exercises and others containing typical problems. Thus, after successfully completing training exercises that contained the most difficult problems, together with realistic time constraints and distractions, the trainee was prepared to successfully complete an integrated simulation of a PAM deploy without aid from an experienced FDO. The philosophy of this PAM Deploy/ICAT (or "PD/ICAT") system was to emulate, to the extent possible, the behavior of an experienced FDO devoting his or her full time and attention to the training of a novice—proposing challenging training scenarios, monitoring and evaluating the actions of the trainee, providing meaningful comments in response to trainee errors, responding to trainee requests for information and hints (if appropriate), and remembering the strengths and weaknesses displayed by the trainee so that appropriate future exercises could be designed. The detailed design of PD/ICAT has been described elsewhere (Loftin, Wang, Baffes, & Hua, 1988, 1989).

Although PD/ICAT has been described as an ITS, it is important to note that its philosophical basis is somewhat different from most ITSs that have been designed for student teaching in school settings. The PD/ICAT system was developed with a clear understanding that training is not the same as

FIG. 9.1. PAM deploy.

teaching or tutoring (Kearsley, 1987). The NASA training environment differs in many ways from an academic teaching environment:

- Assigned tasks are often mission critical, placing the responsibility for lives and property in the hands of those who have been trained.
- Personnel already have significant academic and practical experience to bring to bear on their assigned task.
- Trainees make use of a wide variety of training techniques, ranging from the study of comprehensive training manuals to simulations to actual on-the-job training under the supervision of more experienced personnel.
- Many of the tasks offer considerable freedom in the exact manner in which they may be accomplished.

FDO trainees were well aware of the importance of their jobs and the probable consequences of failure. Although students are often motivated by the fear of receiving a low grade, FDO trainees know that human lives, a billion-dollar Space Shuttle, and the safety of a $100+ million satellite depend on their skill in performing assigned tasks. This means that trainees are highly motivated, but it also imposes on the trainer the responsibility for the accuracy of the training content (i.e., verification of the domain expertise encoded in the system) and the ability of the trainer to correctly evaluate trainee actions. PD/ICAT was intended not to impart basic knowledge of mathematics and physics but to aid the trainee in developing skills for which he or she already has the basic or "theoretical" knowledge. In short, this training system was designed to help a trainee put into practice that which he or she already intellectually understands. The system had to take into account the type of training that both precedes and follows—building on the knowledge gained from training manuals and rule books while preparing the trainee for, and complementing, the on-the-job training that will follow. Perhaps most critical of all, trainees must be allowed to carry out an assigned task by any valid means. Such flexibility was deemed to be essential so that trainees were able to retain and even hone an independence of thought and to develop confidence in their ability to respond to problems, even problems that they never encountered and that their trainers never anticipated.

Development Process

The initial development team consisted of two knowledge engineers and a student. The head of the ODS assigned two Air Force officers, who were scheduled to leave NASA, as subject-matter experts. These two officers had been at NASA for approximately 4 years, had progressed through the traditional training process, had been certified as FDOs, and had successfully executed the task (PAM deploys) in actual Space Shuttle missions. In addition to these experienced subject-matter experts (SMEs), a FDO trainee was also directed to assist and to serve as a test subject for early versions of the system.

The conception and initial planning for the PD/ICAT project began in July 1986, and initial knowledge acquisition was complete by December 1986. The code development began in January 1987, and the system was first demonstrated to the management of the user community in March 1988. During this latter period, three complete versions of the system were developed before PD/ICAT was deemed acceptable by the SMEs. PD/ICAT was delivered for use by both novice FDOs for training and by experienced FDOs for practice and refreshment of skills in March 1988. The rehost of PD/ICAT from the Symbolics machine on which it was first developed to a unix-based workstation was accomplished early in 1989.

The development of PD/ICAT required approximately 3.5 man-years and was accomplished by a mixed team from academia, NASA civil service, and private industry. In addition, a number of students contributed to the project during its life. The actual direct cost of the project was approximately $120,000 (for the academic and private-sector portions). The indirect costs of civil service manpower are more difficult to calculate, but they were approximately $100,000.

Features of the Application

PD/ICAT was composed, in part, of four expert systems that cooperated through and communicated by means of a common blackboard (see Fig. 9.2). This approach was used to permit the segregation of domain-independent knowledge so that the system architecture could be adapted to different training tasks.

Unlike most ITSs, PD/ICAT does not require the trainee to follow a single correct path to the solution of a problem. Rather, a trainee is permitted to select any correct path, as determined by the scenario context. The method used to accomplish this flexibility, without generating a combinatorial explosion of solution paths, is believed to be unique (Loftin et al., 1994).

Error detection occurs through the comparison of the trainee's actions with those of an expert. In the case of complex actions, the error detection is made at the first occurrence to avoid confusing the trainee by detecting all errors that propagated from the one that occurs first.

Error handling is accomplished through the matching of trainee actions with mal-rules containing errors that are commonly made by novices. In

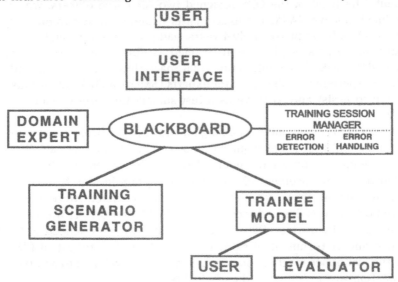

FIG. 9.2. A schematic diagram of the original ICAT architecture used in the construction of PD/ICAT.

addition, the error-handling component decides, based on the trainee model, what type of feedback to give the trainee. Explanations or hints may be detailed for novices and quite terse for more experienced personnel. In some cases, the system may decide not to call attention to the error if there is a reasonable probability that the trainee will catch his or her own mistake.

The training-scenario generator examines the trainee model and creates a unique scenario for each trainee whenever a new session begins. This scenario is built from a database containing a range of typical parameters describing the training context and problems of graded difficulty. Scenarios evolve to greater difficulty as the trainee demonstrates the acquisition of greater skills in solving the training problems.

At the conclusion of each session, the trainee is provided with a formatted trace of the session that highlights the correct and incorrect actions taken, the time required to complete the exercise, and the type of assistance provided by the system. In addition, the trainee's supervisor may view a global history of each trainee's interaction with the system and even generate graphs of trainee performance measured against a number of variables.

Why NASA Should Have Wanted PD/ICAT

Training of astronauts and ground-based flight controllers and system engineers is a massive task. The best training and the mechanism for certifying that personnel have met training objectives occurs through large-scale, integrated simulations. Unfortunately, these simulations require the support of hundreds of people but typically deliver training to only one person in each position. The ability of a given trainee to get significant exposure to a particular process is, therefore, quite limited. The PD/ICAT system, on the other hand, could provide trainees with virtually unlimited access to training in a specific procedure and ensure that the integrated simulation environment could be used to maximum effect.

The End

The original system developed with this architecture (PD/ICAT) has been used by both expert and novice flight controllers at NASA/Johnson Space Center. An extensive investigation of the performance of novices using the system has been conducted. Figures 9.3 and 9.4 show two measures of performance: (a) the time required to perform the nominal task as a function of the number of training experiences, and (b) the number of errors made during the performance of the nominal task as a function of the number of training experiences. It is interesting to note that although the novices used

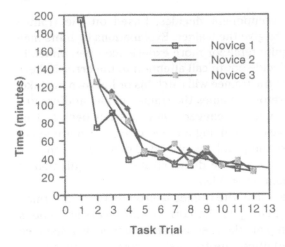

FIG. 9.3. The time required to perform a nominal task as a function of the number of ICAT sessions completed by three novice flight controllers.

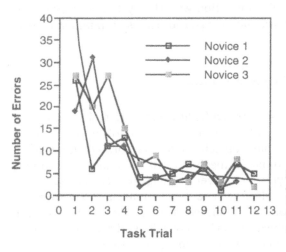

FIG. 9.4. The number of errors committed by novice flight controllers performing a nominal task as a function of the number of ICAT sessions completed.

in this investigation had very different levels of prior experience related to the task, all novices rapidly approached the same level of proficiency.

Just as this ITS reached a useful state and demonstrated its capabilities, the operation for which it was developed was discontinued—an unforeseen and unfortunate circumstance. This "first-of-its-kind" system was, therefore, never used for operational training.

Nonetheless, one could argue that its measured success at meeting its objectives should have led to the operations personnel calling for a full range of ICAT systems to address training of flight controllers for most, if not all, console positions. As noted later, other systems were developed for other NASA centers and for the Johnson Space Center, but none have been requested to support the training of Space Shuttle flight control personnel.

IN THE INTERIM . . .

The initial "success" of ITS technology at the Johnson Space Center stimulated the development of additional ITS applications for use at the Johnson Space Center, the Kennedy Space Center, and the Marshall Space Flight Center. Many of these systems utilize a generalized version of the PD/ICAT architecture that has been patented (Loftin et al., 1994). Following is a brief summary of the specific systems that have been built or are currently under development:

Vacuum Vent Line (VVL)/ICAT

VVL/ICAT was a limited, PC-based, computer-aided ITS for use by mission and payload specialists in learning to perform fault detection, isolation, and reconfiguration (FDIR) on the Spacelab VVL system. This system did not contain the full complement of student modeling, scenario generation, and trainee session management that was described previously. VVL/ICAT provided a reconfigurable schematic of the Space Shuttle's Vacuum Vent Line. As the schematic is altered, the system automatically reflects those alterations in its rule-based description of the VVL system. Users are then trained to perform efficient FDIR of the system by manipulating "valves" and observing the pressures at different points in the system.

Main Propulsion Pneumatics (MPP)/ICAT System

MPP/ICAT is a comprehensive, computer-aided ITS for use by test engineers at NASA/Kennedy Space Center in learning to perform testing of the Space Shuttle Main Propulsion Pneumatics system. This system utilizes the same ICAT architecture as found in PD/ICAT. The Firing Room console environment is duplicated in the MPP/ICAT interface, and training is provided in carrying out the operations and maintenance instruction pertinent to the 750-psi Helium pneumatics system that controls the Space Shuttle Main Propulsion System. In addition to training engineers in nominal test procedures, MPP/ICAT was ultimately intended to address the development and implementation of test procedures employed when faults are detected.

Instrument Pointing System (IPS)/ICAT System

IPS/ICAT was intended for use by payload and mission specialists at NASA/Johnson Space Center and Marshall Space Flight Center in learning to utilize the IPS on Spacelab missions. The IPS is a platform used for mounting and pointing astronomical telescopes during the Astro series of Spacelab missions. The system provides a graphical representation of the Space Shuttle aft flight deck, from where one can access interactive, digitized images of relevant control panels as well as the displays used in operating the IPS. IPS/ICAT was designed to train astronauts in the activation, deactivation, and initial pointing of the IPS as well as in the final pointing of one of the instruments mounted on the IPS (the Hopkins Ultraviolet Telescope). The system uses a modification of the ICAT architecture as found in PD/ICAT and MPP/ICAT.

Center Information System Computer Operations (CISCO)/ICAT

CISCO/ICAT addressed the training of mainframe computer operators in the context of the Johnson Space Center's Center Information System. Operators are provided, in a PC/Windows environment, with a console operator display as well as a "map" of the hardware locations. Utilizing these displays, trainees are instructed in standard operations, including power-up, power-down, and initial process load. Console operations and physical interaction with devices are a part of the training regimen. CISCO/ICAT uses the same ICAT architecture that is found in PD/ICAT and MPP/ICAT.

Active Thermal Control System (ATCS)/ICAT

ATCS/ICAT was designed to train both crew and ground-based flight controllers in the nominal operation and in malfunction recovery of the Space Station Freedom's (a previous design that has been superseded by the International Space Station) thermal control system. This ICAT application is resident on the Apple Macintosh® computer but utilizes the general ICAT architecture.

Object Modeling Technique (OMT)/ICAT

As NASA moves to re-engineer millions of lines of FORTRAN programs, the need to train experienced programmers in object-oriented design is becoming crucial. To this end, OMT/ICAT has been developed to introduce those with backgrounds in procedural program design and development, modern strategies based on the work of Rumbaugh and others. This system employs an architecture similar to that used in the development of IPS/ICAT and has been instantiated on both Windows and Unix platforms.

SUCCESS AT LAST!

The Training Context

During the late 1980s, NASA strongly encouraged the commercialization of Space. One of the outcomes of this program was the creation, by private industry, of a payload known as SpaceHab. SpaceHab is a reusable module that fits into the payload bay of the Space Shuttle and connects, through an airlock, to the Shuttle's middeck. Astronauts are responsible for maintaining both the SpaceHab's systems as well as supervising its experiments. SpaceHab was developed in a extremely short period of time, by NASA standards, at a relatively low cost. Normally, NASA would have developed a Single-System Trainer (SST) to prepare crew members for operating the systems that sustain the SpaceHab module. The time and cost of building a new SST were deemed to be excessive; thus, those responsible for SpaceHab training elected to have an ICAT system take the place of the SST for SpaceHab training. This was the origin of the SpaceHab Intelligent Familiarization Trainer (SHIFT).

Objectives

The SHIFT system was designed to train astronauts and flight controllers in the operation of the SpaceHab module (that first flew on the Space Shuttle in early 1993). The system trains personnel in both the nominal and malfunction recovery procedures for the activation, deactivation, and operation of the SpaceHab subsystems (e.g., electrical, caution and warning, and environmental systems). The system was to be software only and resident on a Sun Sparcstation® workstation.

Development Process

A number of steps were taken to enhance the probability that SHIFT not only meets its intended objectives but also becomes part of the operational training for the first and all following SpaceHab missions. The most important step was the creation of a development team with diverse membership. In addition to the knowledge engineers and software developers, the team consisted of an engineer who had participated in the design of the SpaceHab module, a NASA engineer who was responsible for developing the operational procedures for the module, a member of the training staff who was responsible for delivering and monitoring crew training for SpaceHab, and one of the astronauts assigned to the first SpaceHab mission. The participation of these individuals ensured that the SHIFT system "did its job" and that the intended user played a major role in the system's design.

Features of the Application

SHIFT's interface (see Fig. 9.5) contains a location context aid (upper right-hand corner) that allows the user to "jump" among the four sites where activities occur: the fore and aft flight decks, the middeck, and the SpaceHab module. Within each location, specific controls, panels, and screens can be accessed by selecting them from a view of that location. Upon selection,

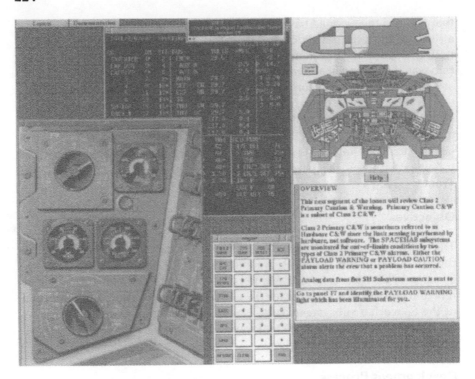

FIG. 9.5. The user interface of the SpaceHab Intelligent Familiarization Trainer
(SHIFT).

each item is replicated (usually in full scale) on the workstation's screen. In
the case of panels, digitized photographs were generally used to insure
realism. The mouse/cursor can be used to change the state of all switches
and controls associated with the tasks to be trained. The lower right-hand
portion of the display is reserved for messages to the trainee.

Trainees can select either a simulation mode that permits free exploration
of over 100 different operational contexts or a training mode that uses a
trainee model to select an appropriate task and level of complexity (basically
determined by the number and nature of anomalies encountered in an
operational context). As trainees carry out actions, the text window echoes
each action. Improper actions are also noted in the window. This information
is the heart of the "after-action review" that each trainee receives at the
conclusion of a training session.

The End

SHIFT was characterized by high motivation on the part of the end users
and strong commitment by all parties during the development process. This
system has been used for four Space Shuttle missions and is now (late 1996)

being employed in training for a fifth; moreover, this system obviated the need to develop an SST (the traditional payload-training approach) for these missions. Thus, this ITS has become the first to enter routine use as a mandatory element of training.

LESSONS LEARNED

From Failure

From the inception of the PD/ICAT project, a number of factors were recognized as essential for success at meeting its objectives and acceptance by its intended audience:

1. The involvement of the ultimate users (or "customers") throughout the development process was judged to be the most important factor to successfully deploy PD/ICAT. The involvement of two subject-matter experts allowed the development team to have at hand the experts in the targeted procedure and to quickly test competing approaches to the solution of specific development problems. However, the most enthusiastic of these experts was transferred from NASA during the project.

2. The subject-matter experts were asked to provide commentary at each stage of the interface development. Thus, at the project's conclusion, the interface had already achieved acceptance by the experts but had never been seen by other "customers."

3. Although detailed documentation was provided to the system's users to enable their own long-term support for PD/ICAT, no training was provided to the intended audience to enable them to provide this support.

4. An important motivation for managers in the intended user community was the ability of PD/ICAT to capture the expertise of personnel who were to be transferred to other areas. This was a key factor in the ready availability of the experts and in management's support of their dedication of time to this project. However, the task for which PD/ICAT was designed was dropped from the Space Shuttle program just as the system was delivered.

To Success

The success of SHIFT has had a major impact on NASA's training plans for the International Space Station (ISS). The Space Flight Training Division, in its original training plans, elected to develop workstation-based simulations of all ISS systems. These simulations were to be supervised by instructors during all phases of use by crews and flight controllers undergoing training. The current plan directs that ICAT technology will be integrated with all of these simulations and used to automate much of this level of training. The focus of effort has now shifted from proving that ICAT technology works to building tools to enable the NASA training community to develop and maintain the ICAT systems that are being deployed for ISS training.

CONCLUSIONS

The initial development of ICAT systems at the Johnson Space Center appeared to be timely, that is, the loss of experienced personnel and the "downtime" occasioned by the Challenger accident offered an ideal opportunity to capture scarce expertise and encapsulate it in workstation-based ICAT systems for flight controllers and, eventually, for astronauts to use for a major portion of their training. The lack of acceptance of the first ICAT system was due both to a loss of its operational target and to a lack of ownership on the part of the intended user community. The success of SHIFT and the adoption of ICAT technology for the ISS program were due to a number of factors, including the direct involvement of both training personnel and astronauts in SHIFT development and the significant savings that accrue from the use of ICAT systems in place of traditional simulation-based training for single systems.

ACKNOWLEDGMENTS

The author is grateful to those who gave of their time and expertise to insure that the ICAT systems described in this paper were based on correct procedures and embodied useful features. Special appreciation is direct to key developers, including Lui Wang, Paul Baffes, Grace Hua, and Steve Mueller. Frank Hughes, Chief of the Johnson Space Center Space Flight Training Division is due special thanks for his unflagging support and willingness to invest time and resources in new technologies for NASA's future.

REFERENCES

Kearsley, G. (Ed.). (1987). *Artificial intelligence and instruction*. Reading, MA: Addison-Wesley.

Loftin, R. B., Wang, L., Baffes, P., & Hua, G. (1988). An intelligent training system for space shuttle flight controllers. *Informatics and Telematics, 5*(3), 151.

Loftin, R. B., Wang, L., Baffes, P., & Hua, G. (1989). An intelligent system for training space shuttle flight controllers in satellite deployment procedures. *Machine Mediated Learning, 5*, 41.

Loftin, R. B., Wang, L., Baffes, P., & Hua, G. (1994). *General architecture for intelligent computer-aided training* (U.S. Patent Number 5,311,422).

10

SHERLOCK 2: AN INTELLIGENT TUTORING SYSTEM BUILT ON THE LRDC TUTOR FRAMEWORK

Sandra Katz
Alan Lesgold
Edward Hughes
Daniel Peters
Gary Eggan
Maria Gordin
Linda Greenberg
Learning Research and Development Center
University of Pittsburgh

The **Sherlock** story began about a decade ago when our research and development team responded to the U.S. Air Force's need for an efficient training technology. Budget cutbacks across the armed services, as well as a shrinking pool of enlisted soldiers prompted the Air Force to look to ITSs as a tool for training avionics technicians expediently in the skills needed to do their jobs, namely, diagnosing faults in, and repairing, faulty aircraft and the systems used to maintain them. In addition to promoting these job-specific skills, the hope was that these ITSs provided technicians with general diagnostic skills that enabled them to transfer readily to related job specialties. For example, avionics technicians whose skills in repairing a *manually* controlled diagnostic system have been honed by an ITS would also learn quickly how to repair an *automatically* controlled diagnostic system.

To meet its training objectives, the Air Force established a program whose main outcomes are a *family* of avionics tutors that support training and cross-training in related maintenance job specialties, and, more importantly, a technology for tutor development. The **Sherlock** tutors (**Sherlock**[1] and

---

[1]For more information on **Sherlock**, see Lajoie & Lesgold, 1989; Lesgold, Eggan, Katz, & Rao, 1992; Lesgold, Lajoie, Bunzo, & Eggan, 1992.

Sherlock 2) are the first in a series of tutoring systems that were developed under this program. As discussed in more detail further on, field evaluations of **Sherlock 2** suggest that the tutor technology we have developed will achieve the goal of skill transfer (Gott, Lesgold, & Kane, 1996). Our focus in this chapter is on **Sherlock 2** and, in particular, on the tutoring system framework on which it was built. **Sherlock 2** is currently being used for training at several U.S. Air Force bases, so it provides fertile ground for discussions about ITS design, development, and deployment.

This chapter proceeds as follows: After a brief introduction to the tutoring system, we describe the LRDC Tutor Framework, that provides the core system architecture as well as the basic ITS "services" that it runs on—that is, coaching, domain expertise, simulation of the task domain, data management services, and so on (see Fig. 10.1). We then focus on the two services which we believe are the most critical for making an ITS behave intelligently: simulations of the task domain and of a domain expert, which can model skilled behavior for students and work with the system's coach to help students acquire domain knowledge and skills. Our discussion of these components of the LRDC Tutor Framework emphasizes the lessons we learned about how to design, develop, and customize the simulation and coaching services for a particular tutor. Finally, we comment more broadly on the lessons we learned during field trials and the early stages of deploying **Sherlock 2** about how to gain and sustain user acceptance of a tutoring system.

OVERVIEW OF SHERLOCK 2

The instructional domain of **Sherlock 2** is F15 avionics. More specifically, its target users are the technicians who work in the intermediate or "back"

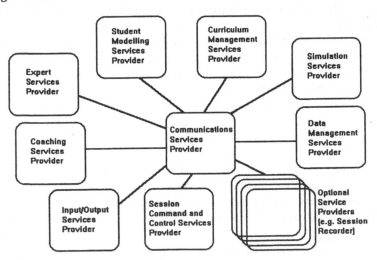

FIG. 10.1. The LRDC tutor framework.

shop, where aircraft modules suspected of malfunction are sent for further inspection and repair by avionics technicians on the flightline. Intermediate shop technicians attach a suspect aircraft module—referred to as a *line replaceable unit* (*LRU*) or *unit under test* (UUT)—onto a manually controlled test station, which is actually a huge switching circuit designed to carry out diagnostic tests. The hard part of the job comes when the UUT checks out, but the test station itself is malfunctioning. It is then the avionics technician's job to troubleshoot the test station to find a faulty component and repair or replace that component. Whereas a UUT contains only a handful of circuit cards, making it fairly simple to isolate an electronic fault, the test station contains about 70 cubic feet of circuitry! **Sherlock 2**'s job is to help technicians acquire the electronic troubleshooting skills needed for this complex diagnostic task.

Sherlock 2 is a realistic computer simulation of the actual job environment. Trainees acquire and practice skills in a context similar to the real setting in which they will be used. The tutor, Sherlock,[2] presents trainees with a series of exercises of increasing difficulty. There are two main phases of a **Sherlock 2** problem: problem solving and review. During problem solving, the student runs a standard set of checkout procedures on a simulation of a test station with a particular UUT attached. Using interactive video with a mouse pointer interface, the student can set switches and adjust knobs and dials on test station drawers, take measurements, and view readings. He[3] can test components by "attaching" probes to measurement points in a video display, replace a suspect component with a shop standard and rerun the failed checkout test, and so on. These and other activities are realized through a menu-driven interface, as shown in Fig. 10.2.

Perhaps most importantly, the student can ask for advice at any point while troubleshooting. Sherlock provides advice at both the circuit path and individual component levels of investigation (see Fig. 10.2). At each level, Sherlock offers functional, strategic, and procedural advice via the *how it works*, *how to test*, and *how to trace* options, respectively. The student is then asked to specify the type and amount of advice he wants by using a submenu of advice options.[4] The basic idea behind this coaching scheme is

---

[2]We use boldface to refer to the tutoring systems, **Sherlock** and **Sherlock 2**, and regular font to refer to the computer coach, Sherlock.

[3]We use the masculine pronoun for simplicity. A significant minority of avionics technicians are female.

[4]For example, if the student asks for how to test help at the component level, he will be prompted to choose between receiving a simple summary of the troubleshooting goals he has already achieved (e.g., *You have verified the A1A2A3 relay card's input and output signals.*); a suggested goal to achieve next (e.g., *You should verify the component's data flow signals.*); a goal to achieve plus specific advice about how to achieve it (i.e., the exact type of measurement to take and pins to test), etc.

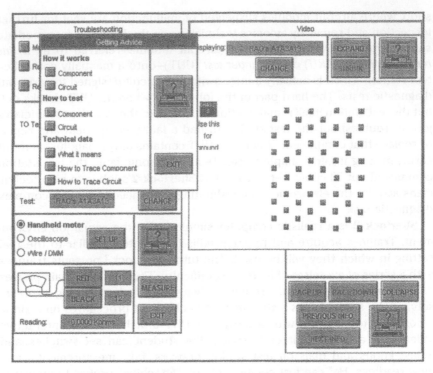

FIG. 10.2. Coaching options in **Sherlock 2**.

to give students control over their own learning and to help them develop
metacognitive skills by requiring them to figure out for themselves what
type of information they need.

In designing **Sherlock 2**, we adopted Vygotsky's (1978) notion of cognitive
tools: "objects provided by the learning environment that permit students
to incorporate new auxiliary methods or symbols into their problem-solving
activity that otherwise would be unavailable" (Derry & Hawkes, 1992, p. 6;
see also Lajoie & Derry, 1993). The major cognitive tool provided by **Sherlock 2** is an intelligent graphics interface that helps students construct a
mental model of the active circuitry and keep track of the status of troubleshooting goals. A sample abstract schematic diagram is shown in Fig.
10.3. Although not visible here, these diagrams are color-coded to indicate
which parts of the circuitry have been found from previous tests to be good
(green), which areas are bad (red), and which are not suspect (yellow) or
of unknown status (black). The drawings are interactive: Selecting any component box produces an explanation of what is known about the status of
that component, given the actions (such as measurements and swaps) carried out so far. An intelligent schematic producer configures these drawings
to match the coach's current explanatory goal, the current problem-solving

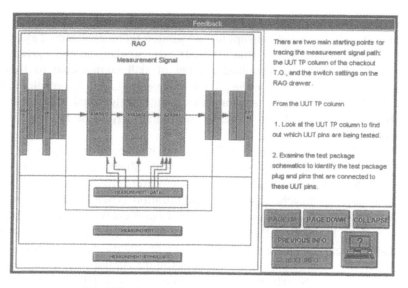

FIG. 10.3. Example of an abstract circuit diagram.

context, and information about the trainee. For example, more space and more expanded circuit detail are provided in the part of the circuit on which an expert might now be focusing or the part the coach wishes to emphasize. Component labeling and color coding are filtered to assure that diagrams don't unintentionally give away too much information.[5]

The design of **Sherlock 2** was also driven by the principles of apprenticeship learning upon which its predecessor, **Sherlock**, was based (Lajoie & Lesgold, 1989), that is, *modeling* of expert troubleshooting behavior, student-initiated *coaching* on the most difficult parts of the task, gradual *fading* of support as expertise is acquired, and review or *reflection* on the student's problem-solving performance, with feedback from Sherlock, the computer coach. Last but by no means least, the system's design was guided by the lessons we learned from developing and evaluating the first-generation prototype (**Sherlock**).

Two fundamental deficiencies in **Sherlock** were corrected in **Sherlock 2.**[6] First, **Sherlock** lacked intelligence, especially in its device modeling and expert performance modeling capabilities. A separate device model was developed for each circuit on which individual problems focused. Similarly, coaching ran on explicitly-written-out hints, encased within the goal structure trees that were hand-coded for each problem. This was not a reusable

---

[5]A component is only labeled if the student has already tested it, if it has been referred to in a coaching message, or if the student has identified the component when prompted. This ensures that students get practice with using schematics to trace through the active circuitry.

[6]These problems are described in more detail in Lesgold, Eggan, Katz, & Rao (1992).

technology by any means. Still, **Sherlock** achieved what we hoped: It provided a dramatic proof-of-concept for automated technical training systems. A controlled study showed that students who worked on the system for approximately 20 hours acquired a level of skill comparable to that of technicians who had been on the job for four years (Lajoie & Lesgold, 1989). This result encouraged us to retain the basic instructional philosophy of **Sherlock** in **Sherlock 2**, but to refine the technology to be intelligent and reusable.

We also felt that the instructional approach itself could be improved. This brings us to the second major difficulty with **Sherlock** that we tried to correct in **Sherlock 2**: compression of all learning activity into the midst of problem solving. Psychological experimentation (Owen & Sweller, 1985; Sweller, 1988; Sweller & Cooper, 1985) and theoretical models of case-based learning (e.g., Mitchell, Keller, & Kedar-Cabelli, 1986) suggest that students often suffer from "cognitive overload" during problem-solving sessions. Consequently, it is best to parcel out some of the instruction to a postproblem reflective phase. In **Sherlock 2**, each problem-solving session is followed by a review phase, which we call *reflective follow-up* (RFU). During RFU, students can step through their solution and receive feedback from the computer coach, replay the simulated expert's solution, review the instructional goals of the tutor, that is, the standards of effective troubleshooting that it is trying to teach—and get suggestions about how to improve in the next session.

Sherlock 2's reflective follow-up is described in more detail later. We turn next to a discussion of the main software development tool that enabled us to develop the flexible, dynamic coaching that is available to students during problem solving and RFU in **Sherlock 2**. This tool also provides a reusable ITS technology, and is called the LRDC Tutor Framework.

THE LRDC TUTOR FRAMEWORK

Motivation

The main motivation for developing our "home brand" of tutor framework was to enable our funding agency, the Air Force, to meet the primary goals of its instructional technology program: to develop a family of avionics training systems and a reusable ITS technology to support this development effort. The framework and its bank of tutoring "services" provided a foundation for developing a set of tutors. A secondary motivation was to make our lives easier. The evolving prototype tutors developed under this program undergo periodic formative evaluations, followed by revisions based on the lessons learned during these evaluations. The more flexible and extensible the system architecture, the smoother this iterative test-and-revise process will go.

For example, during **Sherlock** 2's 4-year gestation, we implemented and field-tested several student modeling technologies (Katz, Lesgold, Eggan, & Gordin, 1992a, 1993; Lesgold, Eggan, Katz, & Rao, 1992), several modes of student-tutor interaction, several postproblem review activities (Katz & Lesgold, 1994), and several different tutoring policies (e.g., we varied the amount of coaching available at different times). Thus, we needed a system in which components could be plugged in, pulled out, replaced, or tuned via parameterization or specialization, without adversely affecting other parts of the tutor. An inflexible system meant major software rewrites as opposed to constrained, local changes. The LRDC Tutor Framework (Fig. 10.1) is the main software tool that we developed to ensure flexibility of design and reusability of code.

What Is a Framework?

A framework is a set of cooperating classes that make up a reusable design for a specific class of software, for example, a framework can be geared toward building graphical editors for different domains such as artistic drawing, music composition, and mechanical CAD. A framework is customized to a particular application by creating application-specific subclasses of abstract classes from the framework.

The framework dictates the architecture of the application. It will define the overall structure, its partitioning into classes and objects, the key responsibilities thereof, how the classes and objects collaborate, and the thread of control. A framework predefines the parameters so that the application designer/implementer can concentrate on the specifics of the application. The framework captures the design decisions that are common to its application domain. Frameworks emphasize *design reuse* over code reuse, though a framework will usually include concrete subclasses that can be put to work immediately (Gamma, Helm, Johnson, & Vlissides, 1995).

Why a Framework Rather Than an Authoring Language? Authoring languages provide an environment for rapid development but do not allow software reusability and extensibility. They support a particular paradigm of application development for a very restricted class of applications, without providing many sophisticated services. Frameworks support rapid development, but they do so by providing reusable services that can be extended using the full richness of the language in which they were implemented. This is particularly important in a research and development environment, where the structure of the framework itself is likely to evolve as the need for new services arises, others become obsolete, interactions are modified, and so on, throughout the iterative test-and-revise process.

The LRDC Tutor Framework is based on the metaphor of an ideal teacher. A teacher needs to have expertise in the domain being taught. However, experts are not always good teachers. In addition to having knowledge and skill in the subject matter, a good teacher also knows how to convey this knowledge—to "coach" students in acquiring knowledge and skills. He or she is also able to assess student understanding. The LRDC Tutor Framework contains Smalltalk objects that take on each of these roles (Fig. 10.1). In reality, individual teaching roles may actually be carried out by a *team* of specialists. For example, the "coach" may really be a coaching team consisting of an expert to make inferences based on the student's actions; a student judge to tailor coaching to the student's level of understanding; and a rhetorician to convey these ideas clearly. In addition to these ITS-specific services, the LRDC Tutor Framework consists of more general services such as data and input/output management tools, and optional services such as a student session recorder.

Communication in the LRDC Tutor Framework

Our aim was to provide a technology for ITS development that was general enough to cut across subject domains and flexible enough to be modified readily. The key to flexibility and reusability is *decoupling* of software components. That is, each software component "hides" its information and implementation mechanisms from other components. Objects do not rely on information about *how* other objects do what they do; they simply need to know *what* services other objects can provide and how to request these services. Decoupling supports a client-server model of communication. Message-passing languages such as Smalltalk are well suited for the client-server relationship because they allow different classes of objects, which provide similar services, to respond to the same protocol or set of messages.

At the heart of the LRDC Tutor Framework is the command control module that coordinates the interaction and communication between services. There are two main types of communication: interobject and interprocess communication. These forms of communication are carried out by the communications manager and the tutor coordinator objects, respectively. The communications manager supports three modes of client-server interaction:

1. *The service request.* In this mode of interaction, the client, who needs a service, initiates a service request. For example, in **Sherlock 2**, the interface manager may need to request that a hint be generated about how a particular electronic component (e.g., a relay card) works. So the client, the interface manager, creates a service request named GenerateComponentHint. The

service request is also given the name of the component as a parameter. The request is then given to the communications manager, who looks up the agent responsible for providing this service in its service providers list. The request is then forwarded to the server, who processes the request and generates a new hint. The server then makes the rest of the system aware of this new hint by publishing the result to a "mailing list."

2. *The mailing list.* Unlike the service request, this mode of communication is initiated by the server rather than the client. An agent can broadcast information to the rest of the system by publishing it to other agents who have subscribed to a particular mailing list. Using the aforementioned example, the hint provider publishes its results to the mailing list named COMPONENT HINTS. On publication, subscribers are notified about the new information by the communications manager. For example, it may be that the object originally requesting the hint (the interface manager) is a subscriber; this object then does something with the hint, such as display it. Other objects who have subscribed to the component hints mailing list are notified as well. For example, a trace manager, whose job it is to produce a step-by-step record of the student-tutor interaction, may have registered to listen in on these conversations so as to be able to record them.

The mailing list communication mode in **Sherlock 2** is weak insofar as only one client can listen for a specific event at any given time. There are no formal semantics for allowing more than one client to listen for an event. There are situations in which we would have liked to broadcast an event to multiple clients and have those clients respond in a predefined sequence. For example, optimally, the student modeler listens for an event brought on by a student's action (e.g., the swapping of a card) but reserves judgment on the action until after the simulated expert has also taken note of the action and posted its opinion about it, for example, the expert deems a swap of a component with a shop standard premature if the student had not even tested the component's outputs to find out if the signal was getting through the component.

Interaction dependencies such as those described in the preceding paragraph may be unordered, partially ordered, or totally ordered. In **Sherlock 2**, the tutor coordinator handles the sequencing of information. However, the communications manager should actually do this because it carries out all of the other communication functions. In the tutoring system for flightline avionics that we are currently building based on the LRDC Tutor Framework, we have implemented a richer semantics that allows the tutor coordinator to inform the communications manager about the correct sequencing of information among multiple clients. The communications manager then carries out this sequence of interactions.

3. *Point-to-point communication.* Under this scheme, one agent needs to send messages to another agent and knows its identity. Furthermore, the

client—who initiates the exchange—might expect an extended conversation with the servant. In other words, this is more than a one-shot communication request. For example, the expert may need to query the simulation for information about the state of the equipment. In this case it is much more efficient for the client (expert) to be able to speak to the server (simulation) directly instead of going through the communications manager. Once the client has finished communicating with the server, the channel is closed and destroyed. Additional mechanisms could be put in place so that these conversations can be listened to by other agents, such as the student trace facility.

Sherlock 2 as an Instance of the Framework

Sherlock 2 is the first instantiation of the LRDC Tutor Framework. As with the next generation of tutors being built on this framework, **Sherlock 2** inherits all of the necessary tutor mechanisms provided by the abstract framework (Fig. 10.1). So what gives **Sherlock 2** its distinctive character? We believe that two things work in tandem to do this: *specialization*, and *policy implementation*. **Sherlock 2** was created by specializing some of the abstract framework classes—most notably, the expert reasoning, student modeling, simulation, and coaching classes—to work in the electronics troubleshooting domain. For example, the simulation engine is customized to model a rather esoteric system (the F-15 manual testing station), but it could readily be specialized for television electronics or, with more extensive customization at a higher level of the simulation architecture, a domain other than electronics.

In addition to specialization, implementation of tutoring policies gives **Sherlock 2** its unique flavor. For example, once the student has solved a **Sherlock 2** problem, he is automatically placed in the postproblem review phase (RFU). During RFU, the student is unable to carry out measurements and receive some of the types of coaching available while problem solving (see Fig. 10.2). Alternatively, we could have made measurement taking available during RFU, so that a student could explore alternative solution paths in light of newly acquired knowledge. However, as with all **Sherlock 2** instructional policies, this one was set by a team of software designers comprised of the LRDC tutor development team and our Air Force colleagues, who weighed the merits and disadvantages of various approaches. The policy was then implemented via the specialized framework classes.

We next look at two examples of how we specialized the framework in **Sherlock 2**. In particular, we describe the simulation services provider and the coaching services provider, focusing on the policies implemented in these modules and the lessons we learned in the process.

THE SIMULATION SERVICES PROVIDER:
LESSONS LEARNED ABOUT DOMAIN MODELING

The Simulation Services Provider "box" of the LRDC Tutor Framework is filled with a collection of utilities for developing accurate and efficient computer models of electronics domains. There are three types of electronics modeling utilities:

1. *Component models*—that is, models of standard electronic components such as relays, switches, diodes, and so on.
2. *Circuit models*—assemblages of components that carry out specific functions—for example, routing of signals from a UUT to a measurement device such as a handheld meter.
3. *Signal propagation utilities*—for example, propagation of ohms and voltage values through the circuit.

While developing **Sherlock 2**, we used this package to simulate the behavior of the F-15 manual avionics test station. We are currently using many of the same utilities to simulate the F-15 aircraft itself in the tutor that we are developing for flightline avionics, **EagleKeeper**.[7] However, we are taking advantage of hindsight and using these utilities more wisely. We now discuss the main lessons we learned about domain modeling.

Lesson I: Aim for Robustness, Coherence, and Modularity in the Simulation Design

To a large extent, these goals were achieved in **Sherlock 2**. In this section, we discuss what we see as the critical factors.

• *Robustness.* By "robust" we mean *domain independent*. We designed the electronic component objects with sufficient generality that they could be used in various modeling contexts. We also developed a set of high-level tools to support modeling of new types of components. Both efforts have paid off, as we are now using these modeling tools and reusing the **Sherlock** package of electronic component objects in **EagleKeeper**.

• *Coherence.* Coherence is the notion that every agent should handle similar activities in similar ways. That is, system elements should adhere strictly to a set of well-established policies for common system activities such as error recovery, data management, and interactions with other subsystems. For example, in **Sherlock 2**, the data management facility services the whole system; data management is not distributed across subsystems,

---

[7]The **EagleKeeper** tutor for F-15 A and C-shop flightline avionics is an additional instantiation of the LRDC Tutor Framework.

so, there is no support built into the simulation for handling data. Similarly, **Sherlock**'s policy for handling services is to have each facility provide a high-level public interface to its services (via the communications manager) while keeping its low-level activities private.

- *Modularity.* A module is self-contained if it can provide its services independent of other modules. This is especially desirable with respect to the simulation because it means that the device model will be reusable in other contexts. **Sherlock 2**'s simulation is largely self-contained, relying only on a small set of services that must be provided externally. In turn, very little of the rest of the tutor depends directly on the simulation. Instead, other system modules rely on a small, well-defined set of services provided by the simulation's public interface.

- By building things in this way, we have decoupled the simulation from the rest of the system while increasing the system's reusability. On the one hand, our simulation could be replaced by a different one as long as the set of services offered by the new package is the same as that supplied by the current simulation. On the other hand, our simulation could be used in other electronics applications simply by tailoring its set of services to suit the needs of the new application.

Lesson 2: Don't Aim for Completeness in the Simulation. Domain Models Should Be Built to Match the Expected Level of Use

Sherlock 2's simulation was "overdone" in the sense that it modeled the entire electromechanical domain of the tutor. To some extent, completeness is useful in that it enables students to experiment freely with the system being modeled. In **Sherlock 2**, students can essentially teach themselves how the test station works, in discovery-learning fashion, by taking various measurements and seeing what happens. Taken a step further, the same package could be used, largely as is, within a job aid that could be set up in the repair shop to allow technicians to verify the values of measurements.

Despite these potential benefits of thorough domain modeling, it was unnecessary for our purposes, as we suspect it is in most cases. Many components—particularly circuit cards such as relay and logic cards—are modeled at an extremely fine level of detail. However, only a portion of these models is actually *used* by the system. Excessive modeling is dangerous for several reasons. Mainly, it makes system maintenance considerably more difficult. Fine-grained models generally mean massive amounts of data. This data has to be managed and version control services provided for it. Special facilities also have to be built to examine and modify the data. In addition, model updates during a trainee's session are very slow.

We are addressing both concerns while developing **EagleKeeper**. First, we are primarily modeling the aircraft at the level that is *visible* to students:

basically, systems and subsystems, down to the circuit card level. Internal behavior is being modeled only on an "as needed" basis.[8] The quantity of data associated with any given model has been reduced by an order of magnitude, and model updates are extremely fast. The **EagleKeeper** model also achieves the same level of fidelity as the **Sherlock** model because the amount of *useful* information available from both models is essentially the same. Although the **EagleKeeper** model lacks the generalizability of the **Sherlock** model across electronic systems, it is far more efficient and easily maintained.

Lesson 3: Design and Build Device Modeling Components Along a Clear Set of Boundaries, Being Careful to Separate the Various "Views" of a Component From Each Other

In **Sherlock 2**, several simulation functions are combined in a single component model. For example, **Sherlock**'s Relay Card object models a relay card from several different points of view, namely, the:

- *Electromechanical point of view.* As expected, the Relay Card object correctly simulates the electrical and mechanical behavior of relay cards.
- *Goal structure point of view.* The Relay Card object can also provide the data structures and information that are used by the expert to assess the goal status of testing relay cards.
- *Instructional point of view.* The Relay Card object can generate coaching messages about itself.

Unmanageable complexity results if these views are confounded. Because components play so many different roles, their code structure becomes unnecessarily dense and complicated. Also, because the methods associated with these classes tend to do double or triple duty, they are hard to maintain.

We are avoiding these problems in developing **EagleKeeper** by providing *separate* objects that can be used by the simulation, the expert, and the coach.[9] This approach has several advantages. First, with three separate

---

[8]Actually, this fairly high level of modeling would not have been adequate for **Sherlock 2**, because the main test station drawer simulated (the Relay Assembly Group, or RAG drawer) is essentially a big switching box. Even so, we did more fine-grained modeling than we should have.

[9]In a similar fashion, the Model/View/Controller (MVC) paradigm of user interface development in Smalltalk decouples the application object (Model), its screen presentation (View), and control over how the user interface reacts to user input (Controller). This decoupling of objects increases flexibility and reuse (Gamma, Helm, Johnson, & Vlissides, 1995).

views of a component, it is hard to couple their behaviors. For example, it is less likely that a software developer makes the expert's view depend on the view used by the simulation. Second, the ability to reuse or replace any given component object is increased. Finally, software maintenance is facilitated because the software engineer responsible for maintaining simulation behavior, for example, is not necessarily the same one who maintains the expert's view or the coach's view.

THE COACHING SERVICES PROVIDER: LESSONS LEARNED ABOUT HOLDING EFFECTIVE TRAINING CONVERSATIONS

Coaching is available while the student is troubleshooting, trying to isolate a fault. Feedback on the student's performance takes place mainly during the postproblem reflective phase (RFU). Both coaching and feedback are the responsibility of the coaching services provider. This module, in turn, relies heavily on the expert model to interpret the student's actions and to suggest alternative actions.

As is true of the other services in the LRDC Tutor Framework, the coaching services provider has been tailored to **Sherlock 2**; it implements particular coaching policies. The main policy is that students should have control over the type of coaching, and the level of detail in that coaching, they receive while problem solving. This policy encourages students to take responsibility for their learning. Sherlock's coaching is noninterventive unless the student does something that is clearly a safety violation (e.g., carrying out an ohms measurement when power is on). This policy of student control carries over into RFU. Students choose the review activities that they want to go through. In this section, we take a closer look at how this policy is carried out during **Sherlock 2**'s problem-solving and review phases.

Coaching During Problem Solving

As shown in Fig. 10.2, there are three main types of advice available to students while they are working on a problem: how to test, how it works, and technical data. The last option provides help with using the various technical documents that avionics technicians commonly work with—for example, tracing through the schematic diagrams of the test station circuitry, following the checkout procedures for a particular UUT, and so on. The coaching messages are handcoded and "static" across problems.

The heart of coaching lies in the how-it-works and how-to-test hints. How-it-works hints convey information about how components and circuit paths carry out the functions for which they were designed. They are dynamic in the sense that operational characteristics change, depending on the equipment configuration, which is, in turn, dictated by the checkout procedure being performed on the test station. Graphic diagrams of the components and circuits that the student requests functional information about are presented along with the coaching text.

The how-to-test hints were designed to help students troubleshoot particular components and circuit paths and the signals that flow through them. These hints recommend troubleshooting goals to try next and—if the student is really stuck—suggest how to achieve these goals. As with how-it-works hints, the how-to-test hints are dynamic because they take into account previous troubleshooting actions and the effects of those actions. These interpretations are reflected in the color-coded circuit diagrams that accompany coaching messages (Fig. 10.3)—for example, red indicates components and signals that are known to be bad, green indicates components and signals known to be good, and so on. These diagrams as well as the coaching messages are updated dynamically to reflect changes in the troubleshooting goal status of the circuitry, and the effects of tests are propagated through the circuit. This provides an explicit, visual interpretation of the effects of student actions. According to user feedback, the circuit diagrams may be the most powerful aspect of **Sherlock 2**'s coaching.

After a student requests a particular type of hint (e.g., how to test) and a particular *level* of advice (e.g., goals remaining—"tell me what goals remain to be solved"), he or she can get increasingly detailed advice in two ways. First, the student can reselect the hint. Second, he or she can request a more detailed hint from the coaching menus. For example, the student who just got advice on which troubleshooting goals remain to be achieved in testing a particular component may not know *how* to carry out these goals. If the student selects how to achieve next goal, he or she will receive a description of a specific measurement to make, including which pin numbers to place the probes on.

Lessons Learned About Developing Coaching Resources

Lesson 1: There Are Pros and Cons to Parceling Advice Across Several Coaching Options and Giving Students Control Over Accessing That Advice. Hypertext Can Help to Eliminate Some of the Problems.

• *Advantages of student control over coaching.* There are several advantages to giving students control over coaching. First, automated representation of student understanding (student modeling) is not yet at the stage where we can reliably determine the kind of information the student needs in most impasse situations; the student probably knows what kind of information is required better than the system. This being the case, maximizing student control over coaching results in the student obtaining needed information without having to wade through unnecessary or distracting text. Perhaps more importantly, there is a greater sense of ownership and control over learning when the student can have some say over coaching rather than letting the tutor make all the decisions.

• *Advantages of organizing coaching by type of information available.* There are advantages to having advice parcelled across several options, according to different types of information, and to offering different levels of detail. For one thing, this approach reduces the amount of text displayed. With limited screen "real estate," and a desire to minimize the amount of text scrolling and paging that users have to endure, finding ways of reducing the amount of text is critical. It also makes it easier for system developers to construct advice that progresses from very general or "reminder" hints to more specific, detailed hints. For example, **Sherlock 2**'s how-to-test coaching levels provide an easy way for students to progress from heavy reliance on coaching—for example, invoking the computer coach to interpret troubleshooting goal states, select the next measurement, and so on—to more independent learning. The how-it-works coaching messages, which progress from generic card function (e.g., relay cards) to specific card function (e.g., test-point-select relay cards) to specific, situated card function (e.g., test-point-select relay cards in the active circuitry)—encourage students to grasp commonalities between components within the same family. This "layered" approach to coaching could prove to be a powerful aid in acquiring the ability to *transfer* skills and knowledge across components, situations, and even job domains.

• *Drawbacks of the Sherlock 2 approach to coaching.* However, various problems also stem from the approach to coaching implemented in **Sherlock 2**. For instance, what if the student does not know what information he or she needs? The student may need to access several hints before receiving useful advice. Also, distributing information over several options makes it difficult for the student to keep track of which type of information is available within each option. For example, the student may continually use how-to-test coaching and be unaware of what how-it-works coaching has to offer. Indeed, analysis of students' use of **Sherlock 2**'s coaching options revealed that students frequently access how-to-test advice and ignore how-it-works advice (Katz, Lesgold, Eggan, & Greenberg, 1996). The reasons for this are unclear. It could be that students do not find the how-it-works hints useful, find the how-to-test hints more memorable because of the color-coded diagrams, are unable to infer what to do from general system information, or some combination of the preceding. Lastly, because the hints are parcelled out, students must continue to request coaching through all of the levels in order to receive the full dose of expertise available if they need it. This has the unfortunate side effect of dividing the relevant information into useful "episodes," while making the "whole story" less explicit. For instance, the full account of *how* to test a component and *why*, based on how the component functions in a nominal circuit, is not available except by requesting both types of hints (how it works *and* how to test) separately. Even if a student does ask for all of the hints available on any particular component or circuit, the burden of synthesizing that information into the full story rests on the student, who may or may not be ready to rise to the task.

• We expect that integration of hypertext with system coaching—as currently implemented in the flightline avionics tutor (**EagleKeeper**)—will help to alleviate these problems with having information dispersed across various advice options. The student will be able to expand general statements as well as select particular terms and phrases in order to receive further information about these items. This also has the nice effect of allowing students to see elaborated explanations in context while limiting the amount of text displayed.

Lesson 2: Get Students Engaged in the Process of Overcoming an Impasse, Beyond Giving Them Control Over the Type and Amount of Information They Receive.

Despite the fact that **Sherlock 2** coaching is primarily noninterventive and students can select the kind (and amount) of information they receive, coaching is, nonetheless, a passive activity in this tutoring system, as it is in most tutors. Sherlock sends a message; the student reads it. Students are not guided in finding the information they need to get over an impasse. In the most extreme case, students catch on that they can "game" the tutor and solve problems simply by repeatedly selecting the most detailed coaching options.[10]

In marked contrast to Sherlock's style of coaching, studies of human tutoring suggest that it is a highly *interactive* process. Experienced human tutors *engage* students in the coaching process through an interaction that

---

[10]A preliminary analysis of system coaching usage, done by our colleagues at Armstrong Laboratory, Brooks Air Force Base (Hall & Rowe, personal communication, 1994), shows that this is indeed what happened in at least some cases. A tabulation of coaching options selected by 21 subjects, during 324 sessions, reveals that at least some students take the path of least resistance. The most heavily used option was **how to test** the circuitry/**how to carry out next goal**, which essentially tells students where to place the probes next. Requests for a goal to try next, without saying how to carry it out, came in second. The least selected options were the more conceptual ones; in particular, the **how it works** hints. Even though knowing how a component (or the circuit as a whole) works can help one to figure out how to test it, students don't seem to want to deal with this degree of indirectness; they want to get right to the punchline! Subsequent analysis of collaborating students' use of **Sherlock 2**'s coaching resources support these observations. Katz's (1997) study of students' interactions with peers and human mentors suggests that "guidance in what to do" conversations dominate problem solving. However, post-problem review is the forum for learning conversations; it is then that conceptual knowledge about the domain is discussed, sometimes in great detail.

Merrill and his colleagues call "collaborative error repair" (Merrill, Reiser, Ranney, & Trafton, 1992). Human tutors prompt students with questions, whose answers will help them to overcome an impasse. One Air Force training monitor captured the essence of collaborative error repair in an informal critique of the tutor: "I don't think the student should be allowed to repeatedly select the next preferred test and just be told what it is and how to do it. It's okay for the student to get some real information the first couple of times, but for me, a real tutor would query the student and make the student tell the tutor what the next preferred test is."

We have been working on a redesign of coaching that will essentially ask students the types of questions that experts ask themselves and, we suspect, that human tutors ask students when they are stuck. For example, the computer coach reminds a struggling student about the measurements he has made so far and prompts him or her to interpret the results of these measurements. Then the coach provides feedback on the student's inter-pretations and prompts him or her to identify an appropriate troubleshoot-ing goal to pursue next. Feedback on the student's reply could again be provided, and so on. Coaching about how the circuitry and its component parts work is available at any time during this process. The effect of this sort of dialogue is to model an expert's means of working through a local planning impasse, that is, deciding what to do next based on what is already known about the circuitry.

Of course, the hardest part about doing "collaborative error repair" well is to strike a happy balance between prompting and informing. We don't want to badger a student with questions when he is asking for help, that is, to answer questions with questions. Providing the student with feedback on the responses and selecting questions according to the student's ability level should help to make a more interactive approach to coaching effective as well as tolerable. Further research on human tutoring is needed to achieve this goal. Our recent work is a step in this direction (Katz, 1997).

Lesson 3: Convey Clearly to Students That While the System's Expert Carries Out a Particular Approach, There Are Others That Might Be Just As Good. Doing So Is Critical to Gaining User Acceptance of the Tutor.

Sherlock's advice represents generally accepted, sound troubleshooting practice. However, nothing that is "generally" sound will be *optimal* in all situations. During field trials of the tutor, we quickly discovered that local shop practices and the idiosyncracies of particular problem situations some-times clashed with the default good advice that Sherlock (the computer coach) was able to provide by consulting with the expert model. For instance, Sherlock recommends that students test components thoroughly before swapping them, to check all secondary data flow inputs to a relay card—whereas shop supervisors at one Air Force base discouraged this practice, deeming it inefficient. Sherlock's advice was sound, especially if viewed from the point of view of parts availability (especially during wartime, swappable parts may not be readily available); however, this advice clashed with locally accepted practice.

Although **Sherlock 2** lost credibility with a few students as a result of shop culture differences such as this, we found that with minimal effort, most students can be brought to understand that the tutor offers one sound approach to troubleshooting, but there are others.[11] We helped to convey this idea by stressing it in our user orientation documents as well as by fine-tuning the language of menu options and coaching text. For example, we changed "Show Sherlock's solution" to "Show a sample expert solution." Unanticipated by us, the former suggested singular correctness, at least to some students. Small details like this sometimes have big effects.

These are just quick fixes to the problems that stem from basing coaching on a single model of expertise. What is really needed are tools that enable users (teachers, shop supervisors, students, etc.) to customize the system expert in accordance with local practices and training goals. For example, a shop supervisor should be able to state whether the system coach is to advise students to check all secondary control signals. Further on, we suggest another approach to addressing this issue, which involves modeling *multiple* domain experts, suitable for coaching at different levels of student ability and in accordance with local preferences. The approach is based on White and Frederiksen's (1987) model progression approach.

Lessons Learned About Developing Post Problem Reflection Resources

The opportunity to review one's performance with feedback from a "master" or coach may be the most valuable aspect of apprenticeship learning situations. Our aim was to capture the flavor of master-apprentice reflective interaction in the tutor. The major design issues we faced were: *What review activities should we provide for students?* and *How should these activities be staged to optimize their instructional value?* In this section, we draw from our experience to present general guidelines on implementing postproblem reflection. Of course empirical research is needed to validate and refine them.

---

[11]In a report on a preliminary user survey conducted by Dr. Ellen Hall and her colleagues at Armstrong Laboratory, Hall reported: "In general, students understand why the tutor teaches the strategy it does and find it useful instructionally" (Hall, personal communication, 1994).

Lesson 1: Limit the Review Activities Available to Students to a Small Set. Let Students Decide Which Activities to Engage in but Suggest One as a Starting Point.

Keeping the Number of Reflective Activities Small. The earliest version of **Sherlock 2** that underwent formative evaluation contained eight review activities, described briefly here. (Menu labels are in parentheses.) Students could:

1. Engage in a step-by-step replay of their solution and receive textual and graphical feedback (the color-coded block diagrams) about what they should have learned about the status of the active circuitry after each action (*Replay my solution*).
2. See a step-by-step replay of an expert's solution (*Replay an expert's solution*).
3. See a side-by-side summary of their actions and the expert's actions (*Compare my solution with an expert's*).
4. View a list of the standards of effective troubleshooting that the coach says the student violated (*Show comments on my solution*).
5. Examine assessment information, namely, how many points they lost by making various types of errors (*Assess my solution*).
6. View graphs depicting their progress on different classes of problems (e.g., problems that involve testing logic cards, problems requiring use of the oscilloscope, etc.) (*Show progress graphs*).
7. Examine their current placement in the curriculum, in which problems are arranged by increasing difficulty, and select a harder or easier problem than the tutor would give them to do next (*Select next problem*).
8. Receive canned advice about the type of coaching they should access in the next session if they are having certain types of difficulties (*Suggestions for next session*).

After the initial field trials of **Sherlock 2**, we reduced the number of review options, in part to make selection easier, but mainly as a result of design changes in the tutor. The final, fielded version of **Sherlock 2** contains five review activities (1-3 and 8 from list, as well as an option that simply lists the standards of effective troubleshooting, called *Review troubleshooting principles*). *Compare my solution with an expert's* appears to be a popular activity, according to a preliminary user's survey conducted by Dr. Ellen Hall at Armstrong Laboratory (Hall, personal communication, 1994). Students enjoy seeing their solution and the sample expert solution side by side, on the same screen, and being able to *select* actions to receive feedback on rather than receive feedback on *all* actions, as in *Replay my solution*.

Selecting a Default Option. As with coaching during problem solving, we parcelled postproblem feedback into several review options to avoid presenting students with screens overloaded with information. We gave students free reign over selecting options and in what order, according to their ability level and experience with the system. For example, a more advanced student may only want to look at an expert's actions, whereas a novice may want to replay each step and ask for comments on every action.

However, during early field trials of the tutor, we discovered that students need more direction than this. Too much choice can be confusing; several participants in the formative evaluation had a hard time deciding which options to select. We, therefore, decided to select one option as the default initial activity and then let the students choose to go through more activities if they so desired.

We chose *Compare my solution with an expert's* as the default initial activity, not because we thought it was the most instructionally valuable but because the activity that we deemed most helpful (*Replay my solution*) ran too slowly. The system has to reinitialize and dynamically reconstruct the goal status of the circuitry during the step-by-step replay, and this takes a bit longer than most students have patience to endure. A slow activity is an unpopular and unused activity. We, therefore, recommend choosing a more efficient one as the default starting point for reflection.

Lesson 2: In Addition to Giving Users Control Over the Types of Review Activities They Engage in, Give Them Some Choice in How to Carry Out These Activities.

In addition to running slowly, *Replay my solution* was tedious because we were asked to implement a policy in which students were required to do a self-critique of each step in their solution. However, even without the self-critique overlay, this activity was taxing for all but perhaps the most novice learners, who might benefit from a slow, careful examination of the feedback available on each action.

Although we believe that self critique as well as peer critique are valuable activities to have students engage in, in developing **EagleKeeper** we are putting implementation of these activities on hold until we learn more about how to do them effectively. In particular, we need to find out what kind of support (coaching) students need during these activities, what advice collaborating students can readily provide to their peers, and how a human expert provides support during collaborative problem solving and review, particularly when students are unable to help each other. Our recent study of peer and student-mentor interaction in **Sherlock 2** is shedding light on these issues (Katz, 1995, 1997).

Lesson 3: Avoid Basing Feedback on a Single, Static View of Expertise. Emphasize Different Standards of Expertise, Across Different Ability Levels and Different Work Cultures.

As we noted with respect to coaching, different work cultures have different standards of what's "right" and "wrong," "good" and "bad." As there is no single, indisputable expert model, user acceptability will suffer if a tutoring system implies that there is.

This happened to us during field trials of an earlier version of the tutor that based feedback on a detailed comparison between the student's action and the expert's action at each solution step. The system penalized students every time their action was inconsistent with the "ideal" solution. Although this kind of feedback may be helpful in optimizing an already good solution, it is meaningless to beginners and anathema to students who are told to do things differently by their supervisors (or teachers). At one Air Force base, students complained that Sherlock was inflexible; some refused to use it.

Fortunately, the damage was not irreparable. We found ways to minimize the problem in the remaining time we had to complete the **Sherlock 2** prototype, but we have more sophisticated ideas now about how to avoid it in the future. In the final version of the tutor, we reduced what Sherlock comments about to a small set of troubleshooting principles (about 10) that most, if not all, experts agree on. These principles were derived from an empirical study of experts' scoring policies (Hall & Pokorny, 1991). We were able to ground criticisms of actions in terms of an objective, agreed-on set of standards. Perhaps more importantly, the *effects* of actions, not the actions themselves, become the subject of commentary. That is, instead of suggesting that "doing X was a bad move," we can now say, "Doing X raised the cost of your solution (or made your solution less efficient than it could have been); an expert would have done Y."

Although this approach still seems sound to us, it does not help to achieve the desirable goal of customizing tutors for particular work cultures and varying student abilities. The same lesson applies here as for coaching during problem solving: software routines and an interface that allow teachers (and students) to add standards or instructional goals that they want to emphasize are very useful and a real boost for user acceptance. Also, under this scheme, the cost of various actions could be set locally, thus reinforcing their relative importance in a particular workplace setting or at a particular performance level.

Another possible route to building more flexible, adaptive tutors is inspired by White and Frederiksen's (1987) model progression approach. Different "expert" models could be built to correspond to different levels of emerging competence and to locally accepted views of expertise. Feedback, assessment, and curriculum decisions could then be driven by this progression of models. So, for example, excessive swapping (replacing a suspect component with one known to be working) is not a major topic of discussion with novice technicians who are struggling to acquire a good troubleshooting strategy such as a binary search of the faulty circuit. However, near experts are discouraged to test by swapping because they should focus on minimizing shop costs.

Lesson 4: Don't Confound Feedback and Assessment. Keep Assessment Information (e.g., Scores, Grades) Out of the Training Conversation Except at the Broadest Qualitative Level. Tailor Feedback to the Student's Current Ability Level.

We found that students tend to fixate on quantitative measures of performance (e.g., scores and graphs) when these are made visible to them and, consequently, fail to grasp the lessons that the feedback is trying to convey. Although we still use quantitative assessment to guide system decision making, for example, to decide whether to advance the students to more difficult problems, we hide these measures from students. Currently, feedback is presented in terms of the qualitative measures of skilled troubleshooting practice that the expert policy-capturing studies of Hall and Pokorny (1991) uncovered.

However, this qualitative approach to feedback could be refined further, and we are taking steps toward doing so in the F-15 flightline tutor, **EagleKeeper**. Instead of commenting on *all* of the things that students could do to improve their performance, the coach in **EagleKeeper** focuses on only two or three of the most important things, as determined by a ranking of troubleshooting standards by domain experts. Eventually, we want to take this one step further and base feedback on the students' current ability level and the things that are most comprehensible to them at that level. For example, the coach suggests that near experts test all of the control signals to a component before swapping it but does not bring this up to trainees as thoroughness of testing should not be a main concern at early stages of learning. The effect is that reaching the point of receiving minor comments such as the need to check secondary data signals will be viewed positively, that is, the coach is running out of things to say! Student diagnosis that is driven by a model progression scheme could support such ability-tailored coaching.

Lesson 5: Use the Curriculum Plan as a Source of Feedback, and Let Students Tailor This Plan According to Their Needs.

In the first version of **Sherlock**, all students went through the same fixed sequence of problems. This sequence was developed in accordance with broad models of difficulty, as defined by domain experts. While developing **Sherlock 2**, the Air Force directed us to implement a more individualized problem selection scheme (Katz, Lesgold, Eggan, & Gordin., 1992b). Although the original approach of having students move through the same sequence of problems provided openings for cheating (e.g., "In problem 3, the A1A2A3 relay card is faulted."), it had other more desirable effects. First, it was computationally manageable. Curriculum schemes become computationally dangerous the more flexible/individualized they get. Also, we could be sure that the fixed problem sequence matched expert trainers' model of a sound curriculum. For example, it ensured that secondary data signal problems were not given too early.

We believe that a sensible compromise between a uniform curriculum and one that is highly individualized is to have a fixed-problem sequence, arranged by qualitative "themes" rather than by levels of mastery (e.g., journeyman and expert). Students could still be moved around within those themes, but the system would know where they are and be able to advise them on what they could learn by going to a particular theme if they suddenly get stuck. This does not require machine intelligence. If the tutor speaks clearly enough to the trainees about which skills can be practiced in each problem theme, students can simply decide for themselves. We believe that this training dialogue should take place during the postproblem reflective stage when students become more aware of their strengths and weaknesses.

LESSONS LEARNED FROM DEPLOYING SHERLOCK 2

Having described the LRDC Tutor Framework and the lessons we learned about designing and developing two of its main components (the simulation and coach), we can now take a look at what we have learned about deploying a tutoring system to its target users. Users is a broad term that can encompass several groups of people. **Sherlock 2**'s users are the students who learn from it, the training monitors who supervise use of the tutor, the shop chiefs who decide who will use the tutor and for how long, and the software engineers at one Air Force base who will take over the job of maintaining the system after the first year of deployment. (Our team of software engineers has not yet challenged this arrangement!)

Despite the design errors that we have candidly described in this chapter, **Sherlock 2** proved to be highly effective for training in the target domain (F-15 "backshop" avionics maintenance, using the manual test station) as well as for near and far transfer. We achieved results better than 1 sigma of effect, in field evaluations conducted by our clients, using blind scoring of real-world performance by control groups, those trained with **Sherlock 2** and those already at master levels of expertise. The system improves problem-solving competence not only in performances directly treated by our training systems but also in solving problems involving circuitry not directly presented in training (near transfer) and even with systems with similar underlying principles but different specific components altogether (far transfer). Table 10.1 shows data from field trials, where the measure is a scale of solution quality in realistic difficult troubleshooting tasks derived via policy-capturing techniques from expert blind rankings of problem solutions.

As noted earlier, a particular goal of our training is to shift technicians from blind parts swapping to scientific data gathering and hypothesis testing. Table 10.2 shows the proportion of troubleshooting actions that were data-gathering measurements, both before and after training. The effects in both Tables exceed 1 sigma and show competence levels approaching those of experts with years of on-the-job experience. In addition to these performance gains, technicians were "generally pleased with **Sherlock 2**," (Gott,

TABLE 10.1

Verbal Troubleshooting Tests

TABLE 10.2

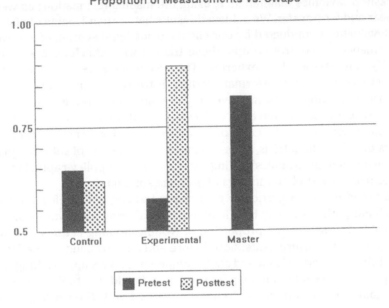

Proportion of Measurements vs. Swaps

Pokorny, Kane, Alley, & Dibble, in press) according to their responses on a written survey ("Tutor Report Card").

Deployment of **Sherlock 2** began in late February 1994. The tutor is now in use at three U.S. Air Force bases. We have learned a lot about what it takes to gain user acceptance and sustain smooth use of an ITS. We present the main lessons as follows.

Lesson 1: Orient Students to the Tutor, but Keep the Introductory Sessions Brief.

In the best of all possible ITS worlds, users would not need an introduction to how to use the tutor, in particular, the features available through its interface. They would be able to "just sit down and use it." We discovered that this was what several users wanted, and there was some disappointment, at least initially, when trainees did not find the tutor completely transparent. It takes relatively little training to get started with **Sherlock 2**. One or two walk-throughs of a problem with a training monitor present seems to be enough for students to be able to do most of the things that they will need to do: set up test equipment, make measurements, access coaching, and so on. Nonetheless, the Air Force would like to eventually do away with the need for training monitors and shop supervisor involvement in computer-based training programs, which means more transparent user interfaces and/or more efficient tutor orientation procedures.

Because the best of all possible ITS worlds is still an imperfect one, users will need at least some amount of orientation to a tutoring system. Ensuring fluent use of the tutor is critical for gaining user acceptance. Frustration arises when students don't know how to do what they want to do. To quote one 7-level user who tried to use the tutor "cold" (without going through the orientation procedure): "Troubleshooting is not hard, but this tutor is." As the training monitor reported to us later, "The student knew what he had to do, but it took about an hour for him to figure out how to do it." In addition to securing user acceptance, a good orientation procedure helps to optimize the tutor's instructional effectiveness. A preliminary survey conducted by Dr. Ellen Hall at Armstrong Laboratory, Brooks Air Force Base, indicated that the small group of students they interviewed were "not making full use of all of the tutor's capabilities. Interestingly, some were not even aware of the assessment criteria used to evaluate their performance until . . . [a] reorientation [was conducted]" (Hall, personal communication, 1994). We are currently analyzing student data to determine usage rates of particular tutor features such as the coaching and review options.

We orient students to the system with the Sherlock 2 Orientation Manual.[12] (In addition, training monitors go through an orientation manual.) The Orientation Manual takes students through three practice problems. The first focuses on the user interface, the second focuses on coaching, and the third introduces the RFU and assessment criteria. Within each problem, students are also told about limitations in the tutor (e.g., constraints on the types of measurements that can be made).

A manual is one approach to tutor orientation, but there are others. The time available for preparation, human resources, and cost are all factors that need to be considered. We started work on an automated introduction to the tutor, but limited time prevented its completion. The new user would have been able to call up an online demo of a **Sherlock 2** problem-solving and review session. The demo would have been interactive, prompting the user to do certain things and to watch the system do the rest. Another possibility we considered was a videotape showing a seasoned user running through a few problems, but production costs were involved and the Air Force had already decided to take the user manual route. It seems to be working satisfactorily, but, again, we need to do more analyses of student data to see which system features students are using.

---

[12]Through its practice problems, the Sherlock 2 Orientation Manual takes a "learning by doing" approach to getting students used to the tutor. Carroll, Smith-Kerker, Ford, and Mazur-Rimetz (1987–1988) argued for this approach in their discussion of the Minimal Manual and demonstrate the effectiveness of this and other features of the Minimal Manual.

Lesson 2: Get User Feedback Early; Get It Often. Most Importantly, Make Sure That Early Users of the System Realize That They Are Instrumental to Its Success.

Well before system development begins, there should be a front-end design review committee that focuses on top-level decisions and sets tutoring policy with respect to the goals of training and the aim of efficient tutor development. At least three concerns need to be addressed by appropriate representatives of this review committee: (a) the funding agency's training and development goals, (b) the specific needs and reactions of users in the field, and (c) scientific review of tutoring methods. Without such a front-end design review and the policies that emerge from it, the tutor development process will fall prey to local or individual preferences, however insightful the latter may be. This happened to us, and it led to frequent, last-minute changes as we tried to accommodate the differing policies of individual bases and even users. **Sherlock 2** became substantially less maintainable as a result of these changes. Although we feel that users, in particular trainers, should be given tools to do some amount of tuning to adjust the tutor to local policies and preferences, we also believe that it is critical for trainers to be aware of the agency's policy on points of contention and for the tutor's default design to reflect these policies.

During formative evaluations of the tutor and in the early stages of deployment (i.e., beta testing), it is absolutely essential for everyone involved to realize that they are an important part of the tutor development process. During field trials of the tutor, our colleagues at the Air Force Human Resources Laboratory, especially Dr. Sherrie Gott, emphasized to all users—shop trainers, training monitors, and trainees—how important their input was to the success of the Air Force's tutor technology development program. This had a huge impact on user interest, involvement, and feedback. In the words of one training monitor:

A point that needs to be understood . . . is that the folks at [Air Force Base X] love **Sherlock** because the Lab made them part of the team. The [Air Force Base X] people were pleased that comments they made about **Sherlock** last summer were incorporated into the tutor. We need to have at least one 7-level [expert] at each base where we put a tutor who understands how important the tutor is and will encourage others at that base to use it. The people I interviewed (and just others who were milling around the office) were really enthused about **Sherlock** because Sgt. X was involved in making **Sherlock** a reality. He even had a meeting on how they would use the tutor on Saturday and no one really complained about the meeting (and that meeting was called knowing they would probably have a recall/alert on the following Monday!).

Since he knows the tutor will help everyone in his shop, everyone wants to use it. They are even willing to overlook the shortfalls and less than user friendly software because of the perceived benefits.

We continue to get user feedback on **Sherlock 2**, although now it is mainly in the less desirable form of software maintenance requests, which we discuss next.

Lesson 3: Centralize System Maintenance.

No matter how much in-house testing goes on before an ITS is deployed, there will still be bugs and features of the tutor that users "just don't like." Maintaining the system is critical for maintaining user acceptance. If a problem remains unfixed, even if the system is still otherwise operational, users will lose interest in it.

We learned this lesson the hard way, soon after **Sherlock 2** had been deployed at two U.S. Air Force bases. A training monitor from one of the bases called to report a fairly serious bug. A recently made software modification had the unfortunate side effect of "confusing" the troubleshooting goal propagation routines. Some goals were marked as satisfied prematurely; others that should have been marked were not. This is a serious problem because the expert looks at the goal status to make strategy decisions and the coach consults with the expert. The result was that Sherlock was giving bad advice, and trainees who experienced the problem were losing confidence in the tutor. The problem was easy to fix, and we did so within a day of receiving the called-in bug report. A new version of the system was immediately shipped to both bases.

The next investigation of usage at the bases, which was conducted by the Air Force, did not take place until about 2 months after this event. The results were surprising. Students seemed to love the tutor at one of the bases. However, students at the other base were ignoring it. It was another few weeks before we found out why. The system upgrade had been installed at the first base but not at the second.

This event reinforced what the Air Force and we had known all along: Software maintenance has to be standardized and centralized. That is, there has to be a standard procedure for reporting problems, for ensuring that they are *real* problems rather than misunderstandings about how to use the tutor, for documenting changes to the system, for distributing upgrades back to the bases, for installing upgrades, and so on. Just as important, there has to be a person or team who is responsible for managing this process and communication between users in the field and the tutor development team. The Air Force's internal software maintenance division devised a procedure that ensured that our team was not constantly badgered with calls from the field and that problems would not only get fixed but also that upgrades would be installed at all bases. Unfortunately, this procedure was not yet fully implemented when the situation described above occurred. This Air Force division, who took over maintenance, acts as the central point of contact between LRDC and the field. When problems are reported, they are first checked by the maintenance division to ensure that they are *real* problems and if they are, a written report called a *Software Problem Report* (SPR, pronounced "spur") is created. SPRs are submitted to us in batches, and we respond with estimated "vital statistics" such as how many man-hours each one will take to complete and a plan for solving each problem. The fixes are incorporated within quarterly upgrades of **Sherlock**, which we submit to the Air Force maintenance division, who then distributes the upgrades to the bases and checks that they have been installed.

This maintenance procedure runs smoothly. We think that this is mainly due to the fact that there is a central group who oversees the process. Whether this group is geographically separated from the software develop-ment team, as in our case, or at the development site is less important than the fact that it exists, somewhere between those who will be making changes to the tutor and those who will be using it.

CONCLUSION

As we have tried to show in this chapter, **Sherlock 2** has many "lessons" to teach ITS developers as well as the students for whom it was built. It sharpened our awareness that there are no hard and fast rules for ITS design, such as "more is better." For example, we discovered that more "views" of an object are better than one when the same object performs several func-tions, but a more detailed simulation of real-world systems is inferior to modeling at the level of use. It heightened our appreciation for the trade-offs that surround almost every design decision, such as whether to parcel information across several coaching options or "package" information within fewer options. It highlighted the value of a tutor framework, especially for developing a family of related tutoring systems. Perhaps most importantly, it stretched our mental model of who the users of an ITS are and who can nourish it into a system that *teaches*, not just one that "works." The toughest part of our job began when **Sherlock 2** met its students and the people who were watching the tutor in action. We feel privileged to have the opportunity to put the "lessons" that **Sherlock 2** and its critics taught us into action in the next generation of avionics tutors we are developing. As with **Sherlock 2** students engaged in post-problem reflection (RFU), its developers were forced to think hard about what our experience meant for future efforts. In the ITS world, as in most realms, hindsight is golden.

ACKNOWLEDGMENTS

The work on **Sherlock 2** discussed in this chapter was funded by the U.S. Air Force and the Spencer Foundation. **Sherlock 2** was a collaborative effort by a team that has included (either currently or in the past) Daniel Abeshouse, Marilyn Bunzo, Linda Greenberg, Edward Hughes, Sandra Katz, Susanne Lajoie, Alan Lesgold, Thomas McGinnis, Rudianto Prabowo, Govinda Rao, and Rose Rosenfeld. Dr. Sherrie Gott, Dr. Ellen Hall, and their colleagues at Air Force Human Resources, Armstrong Laboratories, were active colleagues in the effort. We would also like to thank the editors for helpful comments on an earlier version of the manuscript. None of the collaborators nor the funding agencies necessarily agree with the views expressed.

REFERENCES

Carroll, J. M., Smith-Kerker, P. L., Ford, J. R., & Mazur-Rimetz, S. A. (1987–1988). The minimal manual. *Human-computer interaction, 3*, 123–153.

Derry, S. J., & Hawkes, L. W. (1992). *Toward fuzzy diagnostic assessment of metacognitive knowledge and Growth*. Paper presented at the AERA Annual Meeting, March, San Francisco, CA.

Gamma, E., Helm, R., Johnson, R., & Vlissides, J. (1995). *Design patterns: elements of reusable object-oriented software*. Reading, MA: Addison-Wesley.

Gott, S. P., Lesgold, A., & Kane, R. S. (1996). Tutoring for transfer of technical competence. In B. G. Wilson (Ed.), *Constructivist Learning Environments: Case Studies in Instructional Design* (pp. 33–48). Englewood Cliffs, NJ: Educational Technology Publications.

Gott, S. P., Pokorny, R. A., Kane, R. S., Alley, W. E., & Dibble, E. (in press). *Understanding the acquisition and flexibility of technical expertise: The development and evaluation of an intelligent tutoring system—Sherlock 2*. AL/HR-TR-1997-0014, Manpower and Personnel Research Division, Human Resources Directorate, Armstrong Laboratory, Brooks AFB, TX.

Hall, E. P., & Pokorny, R. A. (1991, September). *Training evaluation in intelligent tutoring systems*. Paper presented at the Fifth Meeting of the DoD Training Technology Technical Group (T2TG), San Diego, CA.

Katz, S. (1995). Identifying the support needed in computer-supported collaborative learning systems. In J. L. Schnase & E. L. Cunnius, *Proceedings of CSCL '95: The First International Conference on Computer Support for Collaborative Learning* (pp. 200–203), Bloomington, Indiana. Hillsdale, NJ: Lawrence Erlbaum Associates.

Katz, S. (1997). *Peer and student-mentor interaction in a computer-based training environment for electronic fault diagnosis* (Tech. rep.). Learning Research and Development Center, University of Pittsburgh.

Katz, S., & Lesgold, A. (1994). Implementing postproblem reflection within coached practice environments. In P. Brusilovsky, S. Dikareva, J. Greer, & V. Petrushin (Eds.), *Proceedings of the East-West International Conference on Computer Technologies in Education* (pp. 125–30), Crimea, Ukraine.

Katz, S., Lesgold, A., Eggan, G., & Gordin, M. (1992a). Approaches to student modeling in the Sherlock tutors. In E. Andre, R. Cohen, W. Graf, B. Kass, C. Paris, & W. Wahlster (Eds.), *Proceedings of the Third International Workshop on User Modeling* (pp. 205–230), Dagstuhl Castle, Germany.

Katz, S., Lesgold, A., Eggan, G., & Gordin, M. (1992b). Self-adjusting curriculum planning in Sherlock 2. *Lecture Notes in Computer Science: Proceedings of the Fourth International Conference on Computers in Learning (ICCAL '92)*. Berlin: Springer-Verlag.

Katz, S., Lesgold, A., Eggan, G., & Gordin, M. (1993). Modeling the student in Sherlock 2. *Journal of Artificial Intelligence in Education* (Special issue on student modeling, G. McCalla & J. Greer, Eds.), *3*, 495–518.

Katz, S., Lesgold, A., Eggan, G., & Greenberg, L. (1996). Towards the design of more effective advisors for learning-by-doing systems. *Proceedings of ITS '96*, Montreal, Quebec. Springer-Verlag Lecture Notes in Computer Science.

Lajoie, S., & Derry, S. (Eds.). (1993). *Computers as cognitive tools*. Hillsdale, NJ: Lawrence Erlbaum Associates.

Lajoie, S. P., & Lesgold, A. (1989). Apprenticeship training in the workplace: computer-coached practice environment as a new form of apprenticeship. *Machine-Mediated Learning, 3*, 7–28.

Lesgold, A., Eggan, G., Katz, S., & Rao, G. (1992). Possibilities for assessment using computer-based apprenticeship environments. In W. Regian & V. Shute (Eds.), *Cognitive approaches to automated instruction* (pp. 49–80). Hillsdale, NJ: Lawrence Erlbaum Associates.

Lesgold, A. M. Lajoie, S. P., Bunzo, M., & Eggan, G. (1992). Sherlock: A coached practice environment for an electronics troubleshooting job. In J. Larkin & R. Chabay (Eds.), *Cognitive approaches to automated instruction* (pp. 49–80). Hillsdale, NJ: Lawrence Erlbaum Associates.

Merrill, D. C., Reiser, B. J., Ranney, M., & Trafton, J. G. (1992). Effective tutoring techniques: A comparison of human tutors and intelligent tutoring systems. *Journal of the Learning Sciences, 2*, 277–305.

Mitchell, T. M., Keller, R. M., & Kedar-Cabelli, S. T. (1986). Explanation-based generalization: A unifying view. *Machine Learning, 1*, 47–80.

Owen, E., & Sweller, J. (1985). What do students learn while solving mathematics problems? *Journal of Educational Psychology, 77*, 272–284.

Sweller, J. (1988). Cognitive load during problem solving: Effects on learning. *Cognitive Science, 12*, 257–285.

Sweller, J., & Cooper, G. (1985). The use of worked examples as a substitute for problem solving in algebra learning. *Cognition and Instruction, 2*, 58–89.

Vygotsky, L. S. (1978). *Mind in society: The development of higher psychological processes*. Cambridge, MA: Harvard University Press.

White, B. Y., & Frederiksen, J. R. (1987). Qualitative models and intelligent learning environments. In R. W. Lawler & M. Yazdani (Eds.), *Artificial intelligence and education: Vol. 1: Learning environments and tutoring systems* (pp. 281–305). Norwood, NJ: Ablex.

11

ARE INTELLIGENT TUTORING SYSTEMS READY FOR THE COMMERCIAL MARKET?

Jeffrey E. Norton
Julie A. Jones
William B. Johnson
Bradley J. Wiederholt[1]
Galaxy Scientific Corporation, Atlanta, Georgia

We have been involved in intelligent tutoring system (ITS) design and development since 1978. In these 19 years, we have developed many successful systems, ranging from basic laboratory research to practical applications of these proven laboratory techniques. Table 11.1 briefly outlines some of these systems, and the Appendix describes each system in further detail.

The term *success* means different things to different people. We feel these systems are successful for several reasons. The first systems focused on basic research (TASK and FAULT), whereas later systems applied the results of these earlier scientific studies. In addition, the systems have received positive results from numerous evaluations. The level of evaluation has varied from a formal, summative evaluation for DG-SIM to usability evaluations for other systems (ATCBI-4, ECS Tutor, etc.). Also, with each new system, we have tried to capitalize on earlier successes, taking the salient features and carrying them forward as well as incrementally advancing the different ITS modules whenever appropriate.

Each of the systems in Table 11.1 adheres to this definition of success. However, even with this success, we must still ask the question: "Are ITSs ready for the commercial market?"

To answer this question, in this chapter, we describe our pragmatic approach to the design and development of these systems that contributed

---

[1]Brad Wiederholt is currently with the Institute for Software Innovation, Inc. in Alpharetta, GA.

TABLE 11.1
ITS Legacy

TUTOR	YEAR	DOMAIN	USERS	SPONSOR	SUMMATIVE EVALUATION
TASK / FAULT	1976-1980	Generic Problem Solving	Laboratory Experiments	NASA & Army Research Institute (ARI)	Yes
SB-3614	1981-1983	Tactical Switchboard Troubleshooting	Signal Soldiers	ARI	Yes
DG-SIM	1984-1985	Diesel Generator Troubleshooting	Utility Maintenance Personnel	Electric Power Research Institute (EPRI)	Yes
Microcomputer Intelligence for Technical Training (MITT)	1987-1993	Space Shuttle Fuel Cell Diagnosis; Missile Message Processing Diagnosis; APU Diagnosis	Astronauts and Flight Controllers; USAF Technicians	NASA; USAF Armstrong Laboratory	No
Environmental Control System (ECS) Tutor	1990-1992	Diagnosis of Aircraft Environmental Control Systems	Airline Technicians	Federal Aviation Administration (FAA)	Yes
MITT Writer	1988-1993	Authoring System for Troubleshooting Tutors	USAF CBT Developers and Instructors	USAF Armstrong Laboratory	N/A
Gas Turbine Information System (GTIS)	1991-1994	Gas Turbine Startup Procedures and Failure Diagnosis	Maintenance Personnel	EPRI	No
Motor-Operated Valve (MOV) Tutor	1992	Motor-Operated Valve Actuator Troubleshooting	Mechanical and Electrical Maintenance Personnel	EPRI; Central Research Institute of Electric Power Industry (CRIEPI)	Yes
ATCBI-4	1991-1993	Air Traffic Control Beacon Interrogator Diagnosis	Experienced Electrical Technicians	FAA	No
Advanced Technology Training System (ATTS) Authoring System	1993-1994	Authoring System for Simulation, Multimedia, Intelligent Tutoring	Utility Instructors and CBT Developers	EPRI; CRIEPI	N/A
F-15 Pneudraulics Tutor	1994-1995	Pneumatic and Hydraulic Systems Troubleshooting	USAF Flightline Technicians	USAF	Yes
MITT Writer for Windows	1995-1996	Authoring System for Troubleshooting Tutors	USAF CBT Developers and Instructors	USAF Systems Program Office (SPO)	N/A

to our success. We then examine how to support a more complete transition of ITSs from the research and development (R&D) environment to the commercial marketplace.

DESIGN AND DEVELOPMENT APPROACHES OF ACCEPTED TUTORING SYSTEMS

For the systems listed in Table 11.1, several common design and development approaches have emerged. These approaches include the following:

- Design for a niche area.
- Keep everyone in the loop.
- Engage the student.
- Develop systems to support users and research.
- Manage the subject matter.
- Push for evaluation.

Design for a Niche Area

Training systems cannot be all things for all people. Systems that claim to be broad based and generic often fall short of having enough specific information to satisfy anyone's training objectives. For that reason, we have always concentrated on a niche for our efforts. The niche has been related to troubleshooting, usually in technical training domains.

Troubleshooting, problem solving, decision making, and fault/medical diagnosis are terms that conjure up descriptions of the most difficult part of any person's job. Be it managers making decisions, technicians troubleshooting an aircraft, or doctors engaging in medical diagnosis, the process requires extensive knowledge and experience. The best in their respective fields are those who can assimilate available information, hypothesize causes, determine alternative solutions or methods to gain further understanding, and then make the diagnosis/decision. Regardless of the occupational domain or industrial application, the demand for good troubleshooting skills is high. Because of this demand, our research and development has focused on the niche related to training for troubleshooting in technical environments.

Our first formal research on computer-based training for problem solving began in the mid-1970s at the University of Illinois (Johnson, 1981; Rouse, 1979a, 1979b). At first, we used a context-free network of interconnected parts to represent fault diagnosis. Problem-solving logic had to be based on the structure of the interconnected parts. Thus, the system was called Troubleshooting by Application of Structural Knowledge (TASK; see the Appendix for a more complete description of TASK). Although the TASK training system predated the term *intelligent tutoring*, the software contained many of the features associated with ITSs. It maintained a transaction file of student actions and provided feedback tailored to each student's actions. The various experimental results derived from extensive laboratory use of TASK were reported by Rouse and Hunt (1984).

Recognizing that real-world problem solving is not context free, the research team designed a context-specific framework to represent real-world component connectivity. The system is called Framework for Aiding the Understanding of Logical Troubleshooting (FAULT; see the Appendix for a more complete description of FAULT). Like TASK, the FAULT system received substantive notoriety in the literature throughout the 1980s (Johnson, 1987; Rouse & Hunt, 1984).

These early research attempts at student modeling, pedagogical advising, and comparing learner performance to optimal performance showed us that there was significant interest and value in diagnostic-related training. That niche had a high payoff for the Department of Defense, other government agencies, and industry. Therefore, we have continued to concentrate on diagnostic-related training, as described throughout this chapter.

Keep Everyone in the Loop

Establishing and maintaining customer buy-in and acceptance is crucial for system success. One way to accomplish this is to keep everyone (managers, end users, subject-matter experts, etc.) aware of and involved with project progress. This can be achieved in several different ways:

- Satisfy the customer (even if there is more than one).
- Employ rapid prototyping techniques.
- Keep grassroots advocates in the field.
- Create tools for the customer to be able to support the software.

Satisfy the Customer (Even if There Is More Than One). Operating in the government R&D arena, contractors must often satisfy more than one customer. Generally, the sponsor is a government laboratory or agency. However, the lab or agency does not normally have the systems on which students must be trained. Instead, the sponsor supports affiliated organizations, as shown here:

Sponsor	End User
Electric Power Research Institute	Participating Utilities
USAF Armstrong Laboratory	USAF User Commands
Federal Aviation Administration	Airline and Aviation Maintenance Schools

However, many times, the sponsor and the end user will have different agendas. Sponsors want to answer specific research questions, whereas end users want to have a system that effectively meets their training needs. Sometimes, these agendas conflict. It is usually possible to satisfy all parties, but the developers must be cognizant of the potential conflict and be prepared to respond accordingly.

With the ECS Tutor (see Appendix), there were really three different customers: the FAA, Delta Air Lines, and Clayton State College (Galaxy Scientific Corporation, 1993a, 1993b). Each of these organizations represented a different part of the aviation industry: regulator, airline/manufacturer, and aviation school. The ECS Tutor was able to answer research questions for the FAA while meeting the different training needs of Delta Air Lines and Clayton State College.

Employ Rapid Prototyping Techniques. Rapid prototyping allows developers to quickly interpret and react to customers' needs. Many times, customers (at all levels) are not sure what they like or dislike until they actually see it (Sewell & Johnson, 1990). Frequent interaction between the developers and the customers provides the customers with immediate feedback that their participation is important to the system's success. Also, by incorporating customer-directed features quickly, the customers feel a strong sense of ownership and pride in the evolving system.

All of the systems listed in Table 11.1 employed different levels of rapid prototyping. The ECS Tutor extensively used rapid prototyping techniques to meet the design requirements of airlines and aviation maintenance schools. The ECS Tutor let the students interact with front panels of different equipment to control the air conditioning on the airplane. Several different areas needed to be validated by the SMEs and the students. Some of these areas included screen layout and navigation, the proper level of fidelity for the air conditioning simulation, the accuracy of schematics and other graphics, the completeness and correctness of supporting materials, and so on. During development, we made numerous iterations through both paper storyboards and software prototypes until the correct "look and feel" were attained. Without prototyping, the final tutor would have only been our interpretation of what the customer wanted.

Keep Grassroots Advocates in the Field. The frequent contact required for rapid prototyping allows the developer to gain grassroots advocates within an organization. Even though it is crucial that a project have management support (for exposure, funding, etc.), it is more important that the project have an advocate at the grassroots level. The advocates may be potential end users, instructors, or instructional developers within the organization.

We have found that grassroots support is particularly important once a system is fielded. With MITT and MITT Writer, user workshops were held during development to get user input (Wiederholt, Norton, Johnson, & Browning, 1992). Air Force personnel from all levels attended the workshops. Comments from the workshops were incorporated into the released version of the software. Once the software was distributed, the people who attended the workshop were strong advocates of MITT and MITT Writer. Once they began to use MITT and MITT Writer, they soon had ideas for enhancements that the developers never imagined. This grassroots support led to a better overall product, as well as follow-on support for additional MITT Writer tutors and projects.

Create Tools for the Customer to Be Able to Support the Software. In today's competitive environment, everyone is looking for ways to reduce costs. The world of ITSs is no exception. After the initial development of the MITT Fuel Cell Tutor, the sponsor was pleased but did not want to have to pay to have every system tailored to its needs. Also, the sponsor knew that the fuel cell itself would change and the tutor would require modifications. Instead of having to pay a third party to make these modifications, the Air Force wanted an authoring system developed. The result was MITT Writer.

When we first developed MITT Writer, we wondered if we had put ourselves out of the tutor-building business by providing the tools for others to develop ITSs (instead of developing the tutors ourselves). Fortunately, the answer was "No." Authoring tools merely accelerated the development process and the rate at which the users wanted enhancements on the development tools and the student run-time side as well. We were then able to pursue future development and application of the technology rather than become a "tutor factory." As a result, a small community of DoD and civilian MITT developers has emerged, complete with user workshops and support for the enhancement of the tool set. In fact, this community has developed over three dozen tutors using MITT Writer, with plans to develop even more.

Engage the Student

The other approaches described in this chapter focus on methodology and infrastructure issues in the execution of a training project. However, looking over the tutors in Table 11.1, we find one common characteristic in the way that they *teach* students: They are fun to use. The systems allow students to become active participants in solving problems, gathering information, controlling simulations, and exploring the environment. Students are given goals and the environment and tools to pursue them. They are allowed to fail; when they do, coaching is available to help them get back on track.

They are allowed to review their progress, and in some cases, compare their progress to that of other students or real experts.

In order to provide systems that are engaging for the student, we have applied three basic features when developing the systems:

- Interactive simulations.
- Guides and advisors.
- Self-monitoring.

Build Interactive Simulations. Interactive simulations allow students to explore realistic representations of their real-world system, gather information, and diagnose malfunctions. In short, simulations allow the student to learn by doing. For example, in the ATTS and MITT genre of tutors (see Appendix), students are presented with a series of real-world views of a system (e.g., control panels, system displays, cockpit layouts, external and internal views of the system, etc). Students are able to wander among these displays. They are occasionally nudged when they are not exploring areas of the simulation that may contain valuable information, but other than that, there is no directed procedure for exploration.

As they wander, students interact with the simulation to perform different functions. For example, in a fuel pump diagnosis tutor built with ATTS, students are able to start and stop the pump, listen for abnormalities, wait a period of time for the pump to heat up, and so on. Actions available in the simulation are based on those available to them in the physical world.

At the beginning of their exploration, students are placed in a situation and given a goal to meet. For example, in a C-130 communications tutor built by the USAF using MITT Writer, students were placed in the role of being a communications specialist during a flight operation. They were given the mission of diagnosing and repairing the broken communications equipment so they could land safely. Though most of the tutors described in this chapter focus on troubleshooting, the same role-playing techniques can apply to other domains as well (such as teaching customer service, inspection and monitoring procedures, or reading comprehension).

The level of fidelity in the simulations can vary from the abstract views of the problem-solving task (e.g., TASK and FAULT) to more realistic multi-media-supported images, videos, device models, and actions available in ATTS. At all levels of fidelity, however, students are able to control and interrogate the system in enough detail to gather enough information to meet their training goal.

Additional interactive features found in some of these systems are available not only when students solve a problem or meet a mission but also before and after these exercises. Many of the systems described are supported by exploratory help manuals in which students can review system diagrams and learn the terminology, operating parameters, and relationships of each of the parts of the device model.

Some of the more recent systems incorporate the ability to review and reflect on the just-completed exercise. During this session, students can review their attempt to solve the problem and compare their actions to those of an expert. The most recent example is the F-15 Pneudraulics Tutor, which combines the best features found in MITT with the best features found in the Basic Job Skills genre of tutors (e.g., Sherlock II). At the end of each exercise, students are allowed to review their progress, trace their actions during the just-completed exercise, and compare their actions to those that an expert takes under identical circumstances.

Provide Guides and Advisors. Many of the systems incorporate some type of advisory function in the learning process. The mixture of interactive simulations and advisors has led to the development of *coached-practice* environments. The basic idea is to allow students to pursue the stated goal or mission of the exercise and when they fail, to step in (if appropriate) and help them get back on track. The systems allow students to learn new facts, concepts, or techniques while they are getting back on their feet and returning to the simulation to solve the problem.

In the systems described here, most of the coaching knowledge comes from either a device model of the system (e.g., cars don't start when the battery is dead; electrical sparks cause certain concentrations of natural gas to explode) or from procedures focused on operating the model (e.g., if the car doesn't start, a cheap thing to do is see if you have gas). In the mid-1980s, we were able to base our advice on functional models such as TASK and FAULT. Over the years, we have extended the techniques to include procedural expertise and more robust device models that can help a student to move from a particular state to the desired one.

Why were TASK and FAULT so widely used both as research tools and as a foundation for subsequent development? The answer is simple—they always had an answer or some advice for the student. These training systems were always able to provide, at a minimum, sound, logical advice regarding troubleshooting. This ability was particularly important because these systems allowed the student to free play the system (an environment in which it is difficult to completely code all branching combinations).

Advice and guidance in the systems are available in both a solicited and an unsolicited manner. Sometimes students decided that they need help; sometimes the system decided. In either case, the students were aware that help tailored to their particular situation was available to them at all times. In some of the tutors, once advice was received, the student was able to browse the advice and help libraries before returning to the exercise. For example, the FAA Environmental Control System Tutor used a four-level help method nicknamed POSH (Procedural, Operational, System, Huh?) that allowed students to browse for information within or among each of these help levels.

Allow Students to Monitor and Evaluate Themselves.

Lastly, another effective method of engaging students is to allow them to monitor and evaluate their progress. Simply by providing the "right" types of information during and after exercises, students begin to take on the responsibility of monitoring their own learning. For example, "right" information includes: showing them their progress over time, highlighting common mistakes made during a variety of exercises, allowing them to choose what to do or learn next, providing reinforcement when progress is being made, showing them what other people have done in similar situations, and so on.

For example, as mentioned earlier, the F-15 Pneudraulics Tutor allows students to review their progress at the end of the exercise and compare their actions to those of an expert. Much of this information is available during the exercise (such as the ability to review actions and mistakes made during that particular exercise). Such reviews can direct students' attention to problems and misconceptions that they are having regarding the system, motivating them to improve.

Sometimes, though, motivation comes through competition between students. For example, with DG-SIM, a score was calculated based on number of errors made, correct steps performed, and time to solve the problem. A list of the top 10 student scores was displayed, and students were told how the scores were calculated. Though just a research project and not part of the regular class curriculum, 2 years of intense battle between students during their breaks and lunches filled up the hard disk with student trace data. We received a quite unexpected service call from the instructor asking us how to fix the system.

Another technique we are beginning to use relates information learned between scenarios or exercises to experiences that occur in real life. In a customer service tutor under development, we integrate computer-based training with the hands-on training that is being conducted in the sponsor's store. The students go through a 3-day period that mixes actual on-the-job experience with that provided by the tutor. While using the tutor, the system refers to the real-life experiences they had the previous day and explains new information in that light. In addition, the tutor refers back to previous computer-based customer service exercises that the student experienced in order to explain a general concept.

The key to engaging and empowering students lies in finding the right level of interactivity and simulation fidelity that lets the students pursue a realistic and worthwhile goal. Advances in desktop computers are making it easier to incorporate these modeling, advisory, and multimedia features into all training applications.

Develop Systems to Support Users *and* Research

We have always been very "product-oriented" with our systems (i.e., we have wanted to leave the end user with a system that is useful in the real world, not just in a laboratory setting). We feel that good science does not have to be at the user's expense. We have been able to maintain this product-oriented approach for the different systems by using the following approaches:

- Refine basic design rather than reinvent the wheel.
- Remain sensitive to costs and schedules.
- Develop and deliver systems on readily available platforms.

Refine Basic Design Rather Than Reinvent the Wheel. Over the years, we have focused our attention on different goals. In the beginning, the emphasis was on rigorous scientific study of methods that were effective in computer-based training for troubleshooting. In the next stage, we developed systems that applied the results of our earlier scientific studies. After developing several systems, we developed a generic training–delivery engine that could be coupled with an authoring system to allow rapid system development and maintenance. Most recently, we have refined the basic design features to meet the needs of a specific domain or to take advantage of improved hardware capabilities and commercial off-the-shelf software tools.

We could have concentrated our research effort on one specific ITS component. However, by doing so, the system would have had limited applicability in the field. Therefore, we chose to make incremental advances in the different ITS components. One example is the evolution of the simulation component of MITT/MITT Writer to ATTS. MITT and MITT Writer employed a surface-level simulation that gave the student sufficient feedback to be able to troubleshoot the problem. However, some systems required a higher level of fidelity than MITT supported.

ATTS expanded the simulation component of the ITS architecture to include a lower level, more robust simulation component (Coonan, Wiederholt, Yasutake, Yoshimura, & Isoda, 1993). The ATTS author uses the system to define components and their behavior. The resulting model, combined with real-time graphical interfaces and multimedia, provides students with an interactive instructional environment with a more detailed representation of the system.

We also saw that evolution does not always mean *adding* functionality. Sometimes, too many features are confusing. Such was the case with the MITT Fuel Cell Tutor (Norton, Wiederholt, & Johnson, 1991). MITT's predecessor, DG-SIM, offered three different types of advice (least cost, most powerful, best test). The users of the MITT Fuel Cell Tutor found this confusing, so the three best tests were merged into one new test—Functional Advice.

Remain Sensitive to Costs and Schedules. One of the driving forces behind the incremental approach described earlier is the constant pressure to reduce costs. It is certainly nice to have an unlimited amount of time and/or money, but that is unrealistic. Therefore, we are constantly looking for ways to give sponsors as much for their research dollars as possible. One simple way is to deliver the system on time and on budget.

Cost and schedule overruns can occur for several reasons. One reason is the lack of well-defined specifications. If there is no firm deliverable, then how do the developers know when they are finished? As a result, developers are always trying to add "one more thing." If developers are not sensitive to costs and schedules, then a project may never make it out of the laboratory.

During the development of MITT Writer, we held two workshops for future MITT Writer developers. The workshops allowed us to evaluate the MITT Writer interface and features. We gained a lot of valuable feedback from the workshop participants and incorporated many of their suggested changes. However, we were under budget and time constraints to deliver the authoring system. Although we made many of the enhancements, we had to establish a cutoff point for the improvements so that we could ensure that the system was delivered on time. MITT Writer was well received when it was delivered on time and on budget to the Air Force. Because we delivered what we promised, we were fortunate enough to receive additional funding to enhance the product.

Develop and Deliver on Readily Available Platforms. One of the keys to fielding a system is to have the system run on a platform that is readily available to the end user. Although specialized AI platforms are excellent research tools, they are highly impractical for the field. Many of our sponsors have been in the government sector. In this arena, procurement of new computer platforms is a slow process. It may take months (if not years) for the newest hardware platforms to make their way down to people in the field. We are aware that many projects sit on a shelf due to developers' lack of sensitivity to the delivery/end-user hardware. Therefore, our designs capitalize on hardware that people can use _now_, not at some unknown, distant point in the future.

For example, MITT was specifically designed for Air Force personnel to use with hardware that they had at their base, not at some special training center. MITT was originally written to run with EGA graphics and 640K of memory (Norton et al., 1991) even though more advanced hardware options existed.

Manage the Subject Matter

Managing the subject matter is a major factor in successful ITS development. Most often, this effort is referred to as knowledge acquisition or knowledge engineering, whereby the subject matter or domain content for the tutor is obtained and organized. Unless the ITS developer happens to also be a subject-matter expert (SME), an ITS project relies heavily on the availability and cooperation of one or more SMEs.

The knowledge acquisition bottleneck can cause a problem for ITS development. However, we have learned the following lessons that can help team members more effectively manage the subject matter:

- Use available resources to shorten learning curve.
- Stay within scope of knowledge needed for the tutor.
- Manage the subject-matter experts.

Use Available Resources to Shorten Learning Curve. There is always a learning curve involved when the knowledge engineer (KE) is not a subject-matter expert. The steepness of the learning curve can be lessened if the KE has some technical background in the general area of the domain. For example, if you are tasked with building a tutor for troubleshooting specialized electronic equipment, a solid electrical engineering background provides a good foundation for learning the specific system. If you do not happen to have an electrical engineer on staff, you may be able to hire one as a consultant to the project. In most cases, however, it is not practical or possible to hire such a person as the KE for each tutor development project.

So what do you do when the KE knows very little about the target domain? To tackle this steep learning curve in a reasonable amount of time, it is important to take advantage of all available resources for learning the domain. One of the first things we typically do is to make an initial visit to the cooperating site. Much can be learned on a tour of the facility where one actually sees the environment and equipment involved in the task. During this initial visit, reference materials such as manuals and technical orders, which can be used off-line for more in-depth study, can be collected. When possible, photographs or videos of equipment that can be used in the final tutor are also collected.

One resource that may be overlooked is the training department within the organization that is involved in the project. This training group may be able to provide training materials that are already prepared to teach the target audience for the tutor. In developing the ATCBI-4 Tutor, the researchers were able to obtain copies of the course text and a schematic book from the FAA Academy. These materials were converted to digital format and included in the training system for online reference.

Also, the training organization may provide a cooperating SME. SMEs who know how to teach, in addition to knowing the domain knowledge, can facilitate the learning process for the KE. In some cases, it may even be possible for the KE to attend a course that is offered by the training organization. In the early stages of developing the MOV tutor for EPRI, the instructors from one of the participating utilities provided a private, condensed course on both the electrical and mechanical portions of the system for the two KEs working on the tutor development. The course included hands-on lab sessions. Without this course, many more hours of independent study would have been needed to learn the equivalent amount of information on the MOV system.

Stay Within Scope of Knowledge Needed for the Tutor. When faced with learning a complex domain for tutor development, a novice KE may get bogged down by trying to learn more than necessary. It is important to remember that the KE's job is NOT to learn everything that the expert knows—it may have taken the expert anywhere from 5 to 30 years to accumulate this level of knowledge. The KE only needs to understand only what information is needed to complete the tutor.

By staying within the niche of technical troubleshooting, we have a decided advantage over those who are starting from scratch on each tutor. Over time, the researchers have been able to define and structure the types of knowledge needed for such a tutor. Eventually, this structure was formally encoded into authoring systems such as MITT Writer and ATTS. These authoring systems reduce the types of information that authors must supply, allowing the author or KE to concentrate solely on the collection of essential data.

For example, in our *conceptual* approach to technical troubleshooting training, we have constrained the knowledge-capturing effort. The tutors focus on the higher level decision-making aspects of learning *what* fault isolation tests to perform rather than on the physical aspects of *how* the tests are performed. Therefore, the tutor does not capture all of the detailed steps on how to complete a particular test. In the knowledge-acquisition stage, the KE can, and should, concentrate on collecting procedural information at the higher conceptual level instead of at the detailed physical level.

By definition, ITSs require an expert model. Even though this model can take a number of different forms, procedural, rule-based expert models are commonly used. However, "complete" expert models (where the expert always provides assistance) are notoriously difficult to develop. To reduce the development costs and increase expert coverage, our ITSs also include a generic expert that generates advice based on the functional connectivity of parts (developed during the earlier TASK and FAULT research).

Manage the Subject-Matter Experts. Generally, experts are very busy individuals. They typically are in high demand due to many years of experience and expertise. Yet, for an ITS project, access to the expert is critical to the development of the expert module. The question of what constitutes an "expert" is somewhat debatable, but we have typically taken a programmatic view of this topic. Project sponsors often determine who the experts are and provide access to one or more of them.

If multiple experts are identified, this raises the question of how to manage knowledge capture from the different sources. One possibility is to use the experts in linear sequence, whereby you do initial capture from one expert and then use the second expert for validation/verification. Although this paradigm may work in some instances, it has potential for much rework if the experts' viewpoints differ drastically. When the ITS for the Space Shuttle Fuel Cell was built, personnel changes meant the linear involvement of two subject-matter experts. Each of the subject-matter experts had a very different mental model of the complex system. The advantage was that the system was ultimately designed to accommodate different perspectives of the same system. However, the obvious disadvantage was increased development time and cost.

An alternative solution that has worked well for us is to use multiple experts in parallel. That is, conduct the knowledge elicitation sessions with both experts at the same time. This approach encourages discussion between the experts, which prompts ideas that may not have occurred by questioning them separately. Also, any disagreements can be worked out prior to implementation. This approach was used for the SB-3614, the Diesel Generator Simulation (DG-SIM), and many of the MITT Tutors. However, this parallel approach becomes unmanageable if the number of experts is expanded (because unanimous agreement among experts is highly unlikely).

If only one expert is available, be aware that the development schedule may be impacted by periods where the SME is unavailable. This situation occurred during the development of the ATCBI-4. We had been working with one SME for a few months when he was sent off to a 6-week training course on a new piece of FAA equipment. Development could not be halted for 6 weeks until he returned. Although there were other individuals available who could support us in the interim, it was apparent that there was considerable time involved in bringing each new person up to speed on what we were trying to do. In this case, we hired a recently retired FAA electronics technician to support them on a part-time basis.

This solution worked very well for several reasons. First, he had access to our office and the domain facility. This made frequent interactions financially and practically feasible. Second, the retired technician had a good working/personal relationship with the initial SMEs; therefore, we did not encounter any acceptance problems when we added him to the project. Perhaps most importantly, he was available on an on-demand basis. Anyone who has done ITS development is familiar with the feast or famine nature of an SME's availability. Whether we had substantial tasks to be done or a single question to be asked, we could get the support we needed. Despite high motivation and interest in a project, too often, SMEs who are working a full day on the job do not have the time to work on the "extra" tasks that they are asked to do to support an ITS development project.

Push for Evaluation

Evaluations of ITSs fall into two main categories: formative and summative. The formative evaluations take place during system design and measure system characteristics such as interface usability and the overall under-standability of the training system to the user population. These aspects of evaluation were described by Maddox and Johnson (1986). Many of the systems described earlier have examined various aspects of formative evaluation.

Although organizations can often find the resources to design and build ITSs, few organizations make the substantial financial commitment to measure the impact of the ITSs with a formal, summative evaluation. Because training organizations are in the business of delivering training in a cost-effective manner, it is understandable that minimal time is spent on controlled, training effectiveness experiments.

As noted in Table 11.1, we have been fortunate enough to have been able to conduct some formal evaluations of our systems. Our evaluations have revealed a number of facts that have contributed to our ITS technology evolution. For example, the TASK and FAULT experiments showed the value of generic problem-solving training. It showed that forced pacing (i.e., rushing the student) did not improve performance. That series of experiments also showed that computer-generated feedback must be clearly explained to maximize learning.

The evaluations with Army trainees on the SB-3614 switchboard showed that effective training could be delivered on low-cost computers as long as the software provided a challenge and good feedback. The DG-SIM evaluations showed that learning retention is greater with computer-based training than with traditional instruction (Maddox, Johnson, & Frey, 1986). Findings such as these required well-planned and controlled evaluations. The MOV Tutor evaluation showed that students trained on the real equipment made a significantly higher number of premature answer attempts as well as more errors. The results suggest that students trained with the MOV Tutor collected more information about a failure before trying to correct the problem (Wiederholt et al., 1993).

DEPLOYING INTELLIGENT TUTORS

ITSs have evolved to a level of sophistication where it is now viable to develop a product that is appealing (and financially attractive) to the commercial sector. However, development for the commercial market is vastly different than the government R&D sector. Developers must adjust to the different expectations of the commercial market and be willing to learn lessons from commercial product developers. In order for ITSs to be commercially viable, developers must consider the following issues—many of which do not always come into play in the R&D environment:

- Determine market demand for the product.
- Adhere strictly to development standards.
- Provide robust suite of development tools.
- Determine marketing strategies for product.
- Prepare to support the product for the long haul.

Determine Market Demand for the Product

One of the major differences between government R&D and the commercial sector is the source of funding. Government dollars await researchers with solid, well-rounded proposals. The selling is limited to convincing someone to fund the idea. In an R&D setting, there normally is limited concern for who would be willing to buy it after development.

However, with commercial ventures, the developers must find someone to fund the project—whether that source is a venture capitalist, an internal R&D budget, or sweat equity from the developers. In addition, developers must consider many different factors that come into play, such as development cost, cash flow, return on investment, and so on. The developers must also show timely and tangible progress or funding could be discontinued. Developers must think of business plans instead of proposals.

In the R&D environment, it is easy to add a particular feature just to see how that feature affects user performance. After all, evaluation of different approaches is the nature of research. Unfortunately, because of cost constraints, additional features must be examined for their utility. Is it a user-driven feature? If not, will the cost of adding the feature be justified by the benefit it provides the user?

Adhere Strictly to Development Standards

Adhering to strict development standards is always a good practice, but it is crucial when developing for the commercial environment. In order to control costs, the developers must develop very clear specifications and tightly follow these specifications during development. Otherwise, as additional features creep in, the development cost will escalate, and the developers will soon find themselves over budget and behind schedule.

The developers must develop and adhere to a rigid test plan in order to ensure software integrity. In the R&D world, systems do not have to be 100% bulletproof as long as they are robust enough to collect the needed data. Many times, features are added at the last minute, without time for adequate testing. However, for long-term success, the commercial product must be as bug free as possible in order to gain acceptance.

Provide Robust Suite of Development Tools

Currently, the cost of custom ITS development is prohibitive in most commercial settings. Many companies cannot afford tens or hundreds of thousands of dollars to have a custom ITS developed. Therefore, the use of authoring systems is a logical and cost-effective alternative. However, given the sophistication of commercial software products, users' expectations are extremely high. Therefore, the authoring system must provide a robust set of development tools to support the author.

Although authoring systems are very useful tools, there is always a trade-off between power and flexibility. In some instances, a general, flexible system is desirable, whereas in other instances, a more specific, powerful system is beneficial. If the authoring system is general, then it can be applied in a wide variety of areas, but it may not be able to support more powerful, specialized functions. Conversely, a very specific authoring system is powerful but loses its general applicability. The developers must decide which type of authoring system is appropriate.

Determine Marketing Strategies for Product

In government R&D, the customer is usually a given—normally the funding agency or an associated user command. However, with a commercial product, the developers must answer several questions about marketing their product (who, what, where, when, and how).

Before developing the product, the developers have hopefully determined who the end user is for their product (what). In addition, the developers must determine whom they will use to reach the audience (who)—whether they become their own sales force or whether they identify an outside party to market their product. They must also determine the appropriate place and techniques to market their product (how and where) and any timing considerations (when)—seasonality, trade show dates, and so on.

Prepare to Support the Product for the Long Haul

Many times, after an R&D project is finished, the ITS lives for a relatively short time. Once its immediate useful life expires, it moves to a shelf in the corner of the lab. However, with commercial products, user needs constantly evolve and mature. The developer must be ready to respond to these changing needs or else be prepared to have their system replaced by a competitor's product.

Due to the anticipated longevity of the product, the developers must constantly be looking for ways to enhance their product. They must devise ways to poll their user population in order to determine the most desired enhancements. Also, the developers must develop a technical support infrastructure to handle support calls, log bug reports, make software modifications, and maintain configuration control.

SUMMARY

After nearly 20 years of ITS research and development, we have seen computer-based training evolve from its primitive beginnings to the current robust, adaptive, multimedia simulations. This chapter has discussed design and development approaches that we believe have contributed to this evolution.

As we consider the question "Are Intelligent Tutoring Systems Ready for the Commercial Market?" the answer is a resounding "Yes." Advances in ITS authoring tools are making the development of affordable ITSs a reality. Tools such as MITT and MITT Writer have already proven their value in Air Force applications, so much so that the Air Force converted the DOS-based product to Windows. The ATTS software, in fact, has already been "shrink-wrapped" and distributed to electric power utilities in the United States and Japan.

However, to ensure the commercial success of ITSs, developers must not only further refine their design and development approaches but also realize that the commercial marketplace contains different requirements and constraints than the R&D world. To remain competitive, developers must react to the new circumstances by either adapting or by joining with someone who already operates in that arena.

ITSs have come too far to be remembered only as a fad. By continuing to evolve the development of ITS authoring tools (just as ITSs themselves have evolved), students will begin to demand the ITS-level of sophistication and interaction for all of their CBT.

REFERENCES

Coonan, T., Wiederholt, B., Yasutake, J., Yoshimura, S., & Isoda, H. (1993). Advanced technology training system: Authoring tools for simulation-based training. *Proceedings of 1993 International Conference on Simulators, Modeling, and Training.*

Crowther, E., & Jackson, J. (1994). An intelligent tutoring system for F-15 flightline troubleshooting. *Proceedings of 1994 Interactive Multimedia '94 Conference*, pp. 16–21, sponsored by the Society for Applied Learning Technology. Also in the *Journal of Instruction Delivery Systems* (Winter 1995, pp. 16–22). The Learning Technology Institute, Warrenton, Virginia.

Galaxy Scientific Corporation (1993a). *Human factors in aviation maintenance: Phase three, Volume 1 Progress report* (DOT/FAA/AM-93/15). Washington, DC: FAA Office of Aviation Medicine.

Galaxy Scientific Corporation (1993b). *Human Factors in Aviation Maintenance: Phase Two Progress Report* (DOT/FAA/AM-93/5). Washington, DC: FAA Office of Aviation Medicine.

Johnson, W. B. (1981). Computer simulations for fault diagnosis training: An empirical study of learning transfer from simulation to live system performance (Doctoral dissertation, University of Illinois). *Dissertation Abstracts International, 41*(11) 4625-A. (University Microfilms No. 8108555)

Johnson, W. B. (1987). Development and evaluation of simulation-oriented computer-based instruction for diagnostic training. In W. B. Rouse (Ed.), *Advances in man-machine systems research* (Vol. 3, pp. 99–127). Greenwich, CT: JAI.

Johnson, W. B., & Norton, J. E. (1992). Modeling student performance in diagnostic tasks: A decade of evolution. In V. Shute & W. Regian (Eds.), *Cognitive approaches to automated instruction* (pp. 195–216). Hillsdale, NJ: Lawrence Erlbaum Associates. Also reprinted in *Educational Technology Research and Development*, 404(4), 81–93.

Johnson, W. B., Norton, J. E., Duncan, P. C., & Hunt, R. M. (1988). *Development and demonstration of microcomputer intelligence training (MITT)* (AFHRL-TP-88-8). Brooks Air Force Base, TX: Air Force Human Resources Laboratory.

Jones, J. A., & Jackson, J. (1992). Advanced computer technology for airway facilities maintenance technicians, *Proceedings of the FAA/NASA Advanced Workshop on Artificial Intelligence and Human Factors in Air Traffic Control and Aviation Maintenance.*

Lesgold, A., Eggan, G., Katz, S., & Govinda, R. (1992). Possibilities for assessment using computer-based apprenticeship environments. In V. Shute & W. Regian (Eds.), *Cognitive approaches to automated instruction* (pp. 49–80). Hillsdale, NJ: Lawrence Erlbaum Associates.

Maddox, M. E., & Johnson, W. B. (1986). Can you see it? Can you understand it? Does it work? An evaluation plan for computer-based instruction. *Proceedings of the International Topical Meeting on Advances in Human Factors in Nuclear Power Systems* (pp. 380–389). LaGrange, IL: American Nuclear Society.

Maddox, M. E., Johnson, W. B., & Frey, P. R. (1986). *Diagnostic training for nuclear power plant personnel, Vol. 2: Implementation and evaluation* (EPRI NP-3829-II). Palo Alto, CA: Electric Power Research Institute.

Norton, J. E. (1992). Intelligent simulation for maintenance training. *Proceedings of the Seventh FAA Conference on Human Factors in Aircraft Maintenance and Inspection* (pp. 101–106). Washington, DC: FAA Office of Aviation Medicine. NTIS No. PB93-146975.

Norton, J. E., & Widjaja, T. K. (1996). *Gas turbine information system (gtis) and jump start training for combustion turbines (JuST CT): Final report.* Palo Alto, CA: Electric Power Research Institute.

Norton, J. E., Wiederholt, B. J., & Johnson, W. B. (1991). Microcomputer intelligence for technical training (MITT): The evolution of an intelligent tutoring system. *Proceedings of the 1991 NASA-Air Force Conference on Intelligent Computer-Aided Training.* Houston, TX: L.B. Johnson Space Center.

Rouse, W. B. (1979a). Problem-solving performance of maintenance trainees in a fault diagnosis task. *Human Factors, 21*, 195–203.

Rouse, W. B. (1979b). Problem-solving performance of first semester maintenance trainees in two fault diagnosis tasks. *Human Factors, 21*, 611–618.

Rouse, W. B., & Hunt, R. M. (1984). Human problem solving in fault diagnosis tasks. In W. B. Rouse (Ed.), *Advances in man-machine systems research* (Vol. 1, pp. 195–222). Greenwich, CT: JAI.

Sewell, D. R., & Johnson, W. B. (1990). The effects of rapid prototyping on user behavior in systems design. *Journal of the Washington Academy of Sciences, 80*(2), 71–89.

Wiederholt, B. J., Norton, J. E., Johnson, W. B., & Browning, E. J. (1992). *MITT writer and MITT writer advanced development: developing authoring and training systems for complex technical domains* (AL-TR-1991-0122). Brooks AFB, Texas: Air Force Systems Command.

Wiederholt, B. J., Widjaja, T. K., Yasutake, J. Y., & Isoda, H. (1993). Advanced technology training system on motor-operated valves. *Proceedings of 1993 Conference on Intelligent Computer-Aided Training.*

APPENDIX: SYSTEM DESCRIPTIONS

Advanced Technology Training System (ATTS)
Authoring System

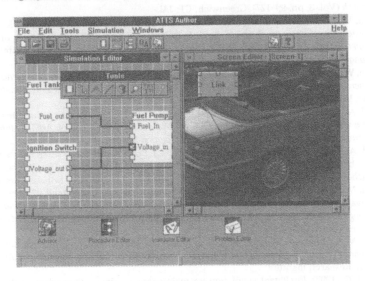

The Advanced Technology Training System (ATTS) is a unified courseware development and delivery environment for technical training (Coonan et al., 1993). The primary goal of the ATTS courseware is to teach maintenance and troubleshooting of complex equipment. The project was conducted under the direction of the Electric Power Research Institute (EPRI) in the United States and the Central Research Institute of Electric Power Industry (CRIEPI) in Japan.

ATTS contains both an authoring system and a student delivery system. The ATTS authoring system allows instructors, subject-matter experts, and training developers to develop and modify training. Instructors can use ATTS as a stand-alone troubleshooting training system or in conjunction with other commercially available training development packages. The

ATTS delivery system merges several key technologies into a unified instructional environment. These technologies include: Interactive Multimedia, Intelligent Tutoring, and Dynamic Simulation.

Using ATTS, authors can combine multimedia (including graphics, sounds, video, animation, and videodisc) to simulate systems and test equipment. To better portray the appearance and behavior of the equipment, authors can create customized controls and readouts. Students interact with these objects to manipulate and test simulated equipment. Authors may introduce failures into simulated equipment; students must locate these failures. Tutoring modules in ATTS guide students through proper troubleshooting procedures and provide advice on logical troubleshooting of faults.

Air Traffic Control Beacon Interrogator (ATCBI-4)

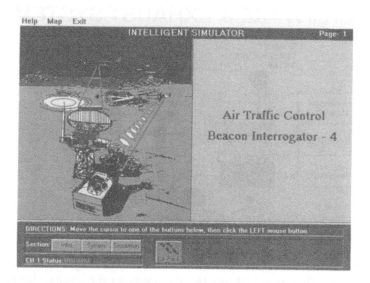

The Air Traffic Control Beacon Interrogator (ATCBI-4) Intelligent Tutor was sponsored by the Advanced Systems Technology Branch of the FAA Technical Center to investigate the use of ITS technology for proficiency training of Airways Facilities electronics technicians (Jones & Jackson, 1992). The ATCBI-4 tutor was designed, developed, evaluated, and refined between November 1991 and July 1993.

The ATCBI-4 is a complex, analog electronics system that is an important component in the Air Traffic Control system. The beacon interrogator augments basic radar data with a unique aircraft identifier and aircraft ground speed. That is, with a working interrogator, an air traffic controller is able to know which radar blip on his display corresponds to a specific aircraft without verbally communicating with the aircraft crew.

Development of the ATCBI-4 Tutor provided challenges in the area of knowledge acquisition and representation. Very little procedural knowledge was documented about how to troubleshoot the system. In addition, there was little documentation on nominal values. Furthermore, the majority of the data used for making troubleshooting decisions was in the form of analog waveforms. Collecting, modeling, and representing the pertinent information for such graphical data was a substantial challenge, given the cost and time constraints of the project.

Diesel Generator (DG) Diagnostic Simulator (DG-SIM)

AIR INTAKE & EXHAUST SYSTEM

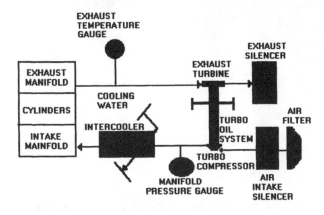

The Diesel Generator (DG) Diagnostic Simulator (DG-SIM), sponsored by the Electric Power Research Institute (EPRI), prepares the student for diagnostic-related job responsibilities on the standby diesel generator. DG-SIM teaches the student to apply a logical approach to troubleshooting for the DG and other plant systems.

While troubleshooting, students are permitted to observe important controls, instruments, and annunciators on the engine local panel. The students can also review functions of each engine system and each component in the schematics section of the simulation. The simulation offers one of four operators' reports for each failure as well as information on test and repair procedures. The students can inspect the engine room and replace parts.

Evaluation of DG-SIM showed that computer-based training has high value with respect to learning retention (Maddox et al., 1986b). In the evaluation, groups were trained with and without the simulation-based DG-SIM. The transfer task was troubleshooting real equipment in the plant. Although there was little performance difference immediately after training, there were significant differences 120 days after training. Those trained with DG-SIM solved problems faster and had fewer errors than those trained with conventional instruction.

Environment Control System (ECS) Tutor

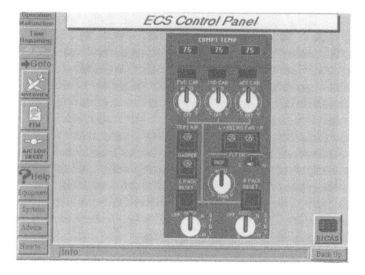

The Environmental Control System (ECS) Tutor, sponsored by the Federal Aviation Administration's Office of Aviation Medicine, provides an interactive environment in which an aviation maintenance technician can diagnose and repair the air-conditioning portion of the ECS of the Boeing 767-300 (Galaxy Scientific Corporation, 1993a, 1993b; Norton, 1992). The tutor combines graphics, animation, and video to give the student the "look and feel" of the real ECS.

The student may operate the Tutor in one of two modes: Malfunction mode or Normal Operating mode. In Normal Operating mode, the Tutor shows how the ECS operates when there is no malfunction. In Malfunction mode, the Tutor groups malfunctions by lesson. Each lesson centers around a common set of symptoms (e.g., one lesson contains all malfunctions that manifest themselves by illuminating the same warning light). After a brief lesson introduction, the student's job is to diagnose and replace the malfunctioning component and then verify proper operation of the system. The ECS Tutor bases its advice on the manufacturer's fault isolation maintenance manual (FIMM). The student may ask for advice at any point in the simulation.

The Tutor not only provides a test bed for simulating ECS failures but also contains declarative knowledge and reference information about the systems and components within the ECS. The student may access several different types of information about components and systems of the ECS.

F-15 Pneudraulics Tutor

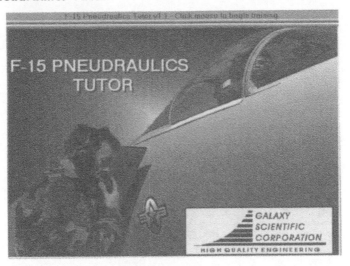

The F-15 Pneudraulics Tutor is a coached-practiced environment to teach troubleshooting techniques in four F-15 pneudraulics systems: Flight Control, Jet Fuel Starter, Hydraulic Power, and Canopy (Crowther & Jackson, 1994). This tutor was funded as part of the U.S. Air Force's Maintenance Skills Tutor (MST) Program, which examines issues related to the rapid and cost-effective development and deployment of ITS technology. All tutor design, development, testing, and training tasks were completed within a 10-month period in 1994 and 1995. In keeping with the intent of the MST program, the F-15 Pneudraulics tutor provides an apprenticeship training environment in which students can develop troubleshooting skills that enable them to solve unfamiliar, cognitively complex troubleshooting scenarios.

Using the F-15 Pneudraulics Tutor, a student progresses through a series of troubleshooting scenarios that pose a simulated system failure, allow the student to walk through the system to make inspections and other diagnostic actions, and identify the failed system component. During this problem-solving process, the student is given instruction on fault and subsystem identification, causal relationships within the affected subsystem, and space-splitting approaches to troubleshooting. At the completion of a scenario, the system provides a set of performance reviews and then selects the next scenario for the student to solve.

The developers based the F-15 Pneudraulics Tutor's development and delivery on previous Air Force projects that created intelligent maintenance tutors. Of particular note, the MITT and MITT Writer projects (described elsewhere in this chapter) served to guide the simulation-oriented, rapid prototyping and authoring approaches to the development of the tutor. The Basic Job Skills (BJS) and Sherlock (Lesgold, Eggan, Katz, & Govinda, 1992) projects served to guide the advanced coaching techniques employed in this tutor.

Framework for Aiding the Understanding of Logical Troubleshooting (FAULT)

TECHNICAL SYSTEM:	Car Engine		Problem 1 of 5	
SYMPTOM:	Cranks But Will Not Start			
GAUGES		**ACTIONS**		**COST**
29 Tachometer Low		Observe (23)Plug wire to (27)Spark Plugs Normal		$1
30 Manifold Pressure		Observe (14)Fuel Pump to (17)Fuel Filter Abnormal		$3
31 Fuel Pressure Abnormal Low		Bench Test (14)Fuel Pump Normal		$8
32 EGT (F)		Observe (13)Coil to (16)Caps Normal		$1
33 Fuel Quantity				
34 Volt Meter Normal				
OPTIONS				
FREE	EXPENSE			
INFORMATION . .	IOBSERVATION . .	O		
COMPARISON . .	CBENCH TEST . . .	. B		
DESCRIPTION . .	[REPLACE	R		
GAUGE	GQUIT	Q		
POSSIBLE . . .P.	P			
ADVICE	A			
HELP	?	YOUR ACTION > 0 13- 16		

Error Message: It was unnecessary to check (13)Coil since you already know that there is Normal Output from the (23)Plug Wires which require Normal Output from the (13)Coil.

FAULT, initially funded by the Army Research Institute, was a research product from the University of Illinois. Unlike the low-fidelity, context-specific simulation called TASK (see description in this Appendix), the FAULT simulation provided real-world representations of technical systems. It provided the basis for nearly 3 years of basic scientific research on learning and has provided a design framework that is used in current ITSs.

The FAULT system permitted the user to obtain information about parts, learn how to perform tests, and then perform the necessary actions to diagnose a malfunction within a system. The FAULT approach has been used for such domains as auto mechanics, aircraft mechanics, military electronics, nuclear power safety systems, space shuttle electrical power systems, and other domains (Johnson, 1987; Johnson, Norton, Duncan, & Hunt, 1988).

The FAULT research helped us to define training system characteristics that were integrated into evolving intelligent tutoring systems. FAULT showed us that the feedback for diagnostic training had to provide specific advice regarding the real-world procedures necessary for success. The research confirmed that practice with a low-fidelity simulation, combined with hands-on equipment practice, provides a very efficient and effective training program.

Gas Turbine Information System

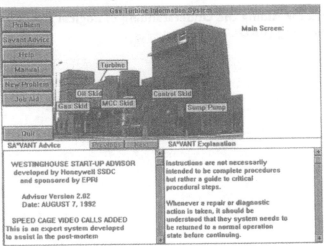

The Gas Turbine Information System (GTIS) was developed for the Electric Power Research Institute (EPRI) (Norton & Widjaja, 1996). GTIS integrates three components: a procedural job aid, an intelligent simulation-based training system, and online manuals.

GTIS simulates a gas turbine maintenance environment that lets technicians troubleshoot simulated turbine faults. GTIS provides the operator the ability to switch at any time between the job aid and the tutor while still allowing the operator to access the online manuals, parts descriptions, equipment descriptions, and diagrams. Thus, the operator can get any additional information about the gas turbine that may be required for troubleshooting or on-the-job aiding. Such context switching allows the operator to learn about the gas turbine without interrupting the job aid.

Microcomputer Intelligence for Technical Training (MITT) and MITT Writer

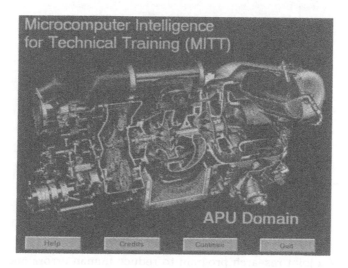

MITT is a training delivery system that specializes in troubleshooting of complex systems (Johnson et al., 1988; Norton et al., 1991). MITT was sponsored by the Air Force's Armstrong Laboratory Human Resources Directorate and RICIS.

MITT uses ITS technology to deliver affordable diagnostic training in a variety of complex technical domains. MITT Writer is an authoring system to create the knowledge used by MITT (Wiederholt et al., 1992). MITT Writer allows the author to design and edit domain-specific files for use by the MITT Tutor.

MITT Writer allows for rapid development and maintenance of a MITT Tutor. It is designed so that the subject-matter expert can also be the developer. Once a new domain is developed, MITT Writer supports easy modification of the domain knowledge. Therefore, as specifications for a system change, the training system can be modified also.

Since 1987, MITT and MITT Writer have been used to build over three dozen tutors in various Air Force User Commands such as Air Education and Training Command (AETC), Air Intelligence Agency (AIA), and Air Combat Command (ACC). Tutors have been developed for a wide variety of domains, including the following systems: Space Shuttle fuel cell, message processing for the Minuteman missile, auxiliary power unit, and high-frequency radio diagnostics. A new version of MITT Writer for Windows provides enhanced pedagogical and multimedia support.

Motor-Operated Valve (MOV) Tutor

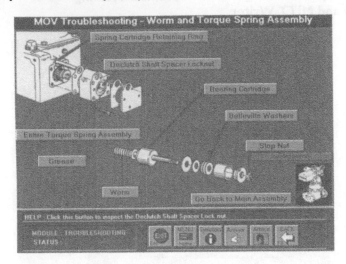

As part of a joint research program to reduce human errors and improve productivity in U.S. and Japanese nuclear power plants, the Electric Power Research Institute (EPRI) in the United States, and the Central Research Institute of Electric Power Industry (CRIEPI) in Japan identified training as an appropriate intervention to enhance troubleshooting proficiency on motor-operated valves (MOVs) (Wiederholt et al., 1993). MOVs are electromechanical devices used throughout nuclear power plants in many different systems. Because these valves are so numerous, systemic problems with any part of an MOV assembly can have negative effects on plant performance.

There are four MOV training modules within the tutor. The first module provides workers with fundamental knowledge in the MOV operation and overhaul task. The second and third modules increase the students' awareness of the mechanical and electrical operation of the MOV actuator. The final module is a troubleshooting simulation that allows students to integrate their knowledge of MOVs, become active participants in diagnosing failures, and solve many different diagnosis problems in a limited amount of training time.

The MOV Tutor was developed by first conducting a task analysis to document training objectives and implementation strategies for the courseware. Next, the courseware was built based on techniques from the MITT research program. Two evaluations were then conducted to measure the courseware's usability and effectiveness. Evaluation of the MOV Tutor showed that the structure and form of the training interface were ideal for MOV students. The students rated the visual interface highly and sufficient for learning the material. The effectiveness evaluation demonstrated the tendency of students trained with the MOV Tutor to perform more inspections and to be more cautious in diagnosing and correcting failures, as shown by the significantly higher accuracy ratings, and higher number of actions performed before attempting to solve troubleshooting problems.

Troubleshooting by Application of Structural Knowledge (TASK)

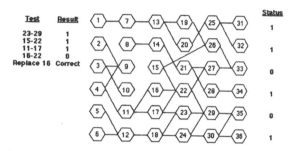

TASK - **T**roubleshooting by **A**pplication of **S**tructural **K**nowledge

Members of our research and development team embarked on computer-based diagnostic training research in the late seventies (Johnson, 1981; Rouse, 1979a, 1979b). Troubleshooting by Application of Structural Knowledge (TASK), first funded by NASA then by the Army Research Institute, was designed as a tool to study how humans collect and process information during problem solving.

TASK graphically displayed a network with a single failed component. Users diagnosed the system by performing tests between components. Over the 5-year experimental research life of TASK, a variety of student-modeling and feedback variations were integrated into the software. By 1979, versions of TASK had many of the features that became necessary attributes of today's ITSs. These features have been reported extensively (Johnson & Norton, 1992; Rouse & Hunt, 1984).

The TASK experiments of the 1970s provided valuable insights on how humans collected data for problem solving and on what kind of feedback best affects learning. TASK showed that feedback must be explained and the quality of the computer-based explanation was directly proportional to learning. This simple principal is a basic foundation of today's ITSs.

Tactical Communications Equipment Trainer (SB-3614)

In the early 1980s the Army Research Institute funded the first effort to transfer the TASK and FAULT research results into actual Army training. The Army SB-3614 Tactical Switchboard training system attempted to bridge the gap between low-fidelity simulation and real equipment (Johnson & Norton, 1992). The system was designed to permit Army students to see graphical depictions of the switchboard. The graphical depictions raised the fidelity of the simulation so the students understood that the simulation-based training had a reasonable expectation of transfer to the real equipment. As an early ITS, the training system tracked student actions, recorded errors, number of actions, and time to completion. Feedback was based on adherence to logical rules of troubleshooting.

AUTHOR INDEX

SUBJECT INDEX

A

Apprenticeship, 75, 87, 207, 231
ATCBI-4 Tutor, 279
ATCS/ICAT, 222
ATTS, 265, 278
Authoring
    authoring environments, 163
    authoring language, 233
    authoring process, 108, 135
    authoring system 128
    authoring tools, 103, 135

B

Behavioral Technology Laboratory,
    137
Blackboard, 218
Black-box method, 151
Bug library, 34, 150
BUGGY, 34
BYGONE, 152

C

Cardiac arrest tutor, 8
CISCO/ICAT, 222
Client-server, 234
Coached practice, 266
Cognitive tools, 230
Collaboration, 15, 189, 191, 197, 207
Collaborative error repair, 244
Collocation, 121
Community of practice, 17
Computer-based training, 4, 5, 54,
    77, 79, 129, 189
Consensus building, 89
Cost effective, 59

Cost-benefits analysis, 51, 53, 60,
    76, 79, 99
Costs
    capital costs, 58
    delivery costs, 54
    development costs, 54
    labor costs, 57
    overhead costs, 54
    productivity costs, 57
    start-up costs, 67
    steady-state costs, 67
    system development costs, 55
    training delivery costs, 56
    training development costs, 54
    training overhead costs, 57
CSCW computer supported
    cooperative work, 203
Cultural differences, 117, 245

D

Decision making, 134
Desktop video conferencing, 204
DG-SIM, 259, 280
Differentiation advantages, 51, 61
DIME, Distributed Intelligent Multi-
    media Environment, 201-207
Distance learning, 189, 190, 205
Domain assessment, 78, 79
Drill and practice, 191

E

Early adopters, 16
ECS Tutor, 263, 281
Emerging constraints, 170
Empirical investigations, 21
Epistemology, 10

Printed and bound by CPI Group (UK) Ltd, Croydon, CR0 4YY

17/10/2024

01775706-0002